159 Anaesthesiologie und Intensivmedizin
Anaesthesiology and Intensive Care Medicine

vormals „Anaesthesiologie und Wiederbelebung"
begründet von R. Frey, F. Kern und O. Mayrhofer

W0050031

Herausgeber:
H. Bergmann · Linz (Schriftleiter)
J. B. Brückner · Berlin   M. Gemperle · Genève
W. F. Henschel · Bremen   O. Mayrhofer · Wien
K. Meßmer · Heidelberg   K. Peter · München

G. Sprotte

Thermographic Investigations into the Physiological Basis of Regional Anaesthesia

With 20 Colour Figures

Springer-Verlag Berlin Heidelberg GmbH 1985

Priv. Doz. Dr. med. Günter Sprotte

Institut für Anaesthesiologie der Universität Würzburg,
Josef-Schneider-Straße 2, D-8700 Würzburg, FR Germany

Translated from the German by

David Roseveare

Rohrbacher Straße 54, D-6900 Heidelberg, FR Germany

ISBN 978-3-540-12638-6   ISBN 978-3-642-69268-0 (eBook)
DOI: 10.1007/978-3-642-69268-0

Library of Congress Cataloging in Publication Data. Sprotte, G. Thermographic investigations into the physiological basis of regional anaesthesia.
(Anaesthesiologie und Intensivmedizin = Anaesthesiology and intensive care
medicine ; 159)
Bibliography: p. 1. Conduction anesthesia – Physiological effect.   2. Anes-
thetics – Physiological effect.   3. Medical thermography.   I. Title.   II.
Series: Anaesthesiologie und Intensivmedizin ; 159.   [DNLM: 1. Anes-
thesia, Conduction.   2. Thermography.   W1 AN103YJ v. 159 / WO 300
S771t] RD84.S67   1985   617'.964   84-18612
ISBN 978-3-540-12638-6(U.S.)

Typesetting: Elsner & Behrens GmbH, Oftersheim

2119/3140-543210

Foreword

The existence of a differential block is still part of the theory of regional anaesthesia. In 1980 it was described in detail by Cousins and Bridenbough in their standard work *Neural Blockade*.

The theory of differential sensitivity of fibres in the peripheral nervous system essentially goes back to Gasser and Erlanger, who in 1929 established that when isolated peripheral nerves are electrically stimulated in the presence of increasing concentrations of cocaine, the compound action potentials of slow-conducting fibres are blocked before those of fast-conducting fibres show any measurable changes.

In man, regional anaesthesia begins subjectively with a feeling of warmth, objectively with a corresponding increase in skin temperature. There is then, in order of occurence, loss of sensation of cold, heat and pain, and pressure and touch, and finally loss of voluntary motoricity. In recovery from anaesthesia, these return in the reverse order.

The theory of differential block is by no means undisputed. In 1981, de Jong, commenting in *Anesthesiology* on the work published in that journal by Gissen et al. which showed a new sequence of differential sensitivity in the rabbit, wrote, "There remains plenty to be done yet before the book on differential nerve block can be closed".

In this monography, my colleague Dr. Sprotte presents the methods and results of his studies on the clinical application of differential block. He began with observations and measurements of skin temperature with the aid of telethermography, finding that in both extradural and peripheral regional anaesthesia, the amount of heat given off by the skin continues to increase until voluntary motoricity is completely blocked. He then worked systematically with differential block in clinical practice. His crucial line of thought is that a reliable statement is possible only if blockade of the sympathetic regulation of vasomotor tone can be recorded quantitatively in analogy to measurement of, for example, voluntary motoricity. He recognized that one cannot speak of a vasomotor blockade when skin temperature increases by only 1° or 1.5 °C.

Dr. Sprotte methodically introduced two innovations. One was continuous perfusion with a local anaesthetic (e.g. in the brachial plexus, the tibial nerve and the common peroneal nerve), the other was stimulation of vasomotor tone by cold. The principal function of the sympathetic innervation of the vessels of the skin in the hands and feet is temperature regulation. The reflexive vasoconstriction reaches its maximum at a skin temperature of 15 °C.

Dr. Sprotte's decisive finding was that in all investigations, the sympathetically innervated vasomotor system in the hands and feet became fully blocked at the same time as or after the voluntary motor system, and almost always together with the sensation of touch. The sensations of cold and pain were consistently blocked before the sensation of touch and the vasomotor and voluntary motor systems. Despite the early extinguishing of the sensation of cold, the function of the cold reflex was not disturbed; icewater cooling led to measurable vasoconstriction right up to the point of total block. The dissociated loss of the various sensations in regional anaesthesia is therefore not due to a clinically appreciable differential sensitivity of fibres in the peripheral nervous system.

I am delighted that Dr. Sprotte's work has been awarded with the Carl Ludwig Schleich Prize of the German Society for Anaesthesiology and Intensive Medicine. I am sure that his monograph will be of equal interest to clinician and theorist alike, and I hope that it will be as widely destributed as possible and that the author receives the recognition he deserves.

Würzburg, im Oktober 1984   K. H. Weis

Contents

Physiology and Pharmacology of Regional Anaesthesia

Mode of Action of Local Anaesthetics

The physiological process of stimulation and conduction of stimuli in the nervous system is based on the specific ability of nerve membranes to change their permeability for sodium ions. This permeability of the membrane is regulated by lipoproteins [59] which are thought to have contractile properties [45]. At rest they seal the membrane against the high extracellular sodium concentrations and thereby guarantee the resting potential. When the nerves are sufficiently stimulated, the lipoproteins suddenly change their structure (probably through a separation of membrane-bound calcium ions) and allow so-called sodium channels to form. This leads to a rapid influx of sodium ions and the resting potential over the membrane breaks down. If this oscillation of potential exceeds a certain level, the mechanism is transmitted to neighbouring sections of membrane and the stimulation is conducted further [12]. Local anaesthetics obstruct stimulation and conduction of stimuli by conserving the membrane's resting potential. Under the influence of these substances the lipoproteins can no longer open any sodium channels [59].

The underlying mechanism has not yet been fully clarified; of the many theories advanced [9, 15, 46, 51, 64], only a few remain current. According to Blaustein and Goldman [5], local anaesthetics and calcium ions compete for reactive positions in the sodium channels that are normally occupied by the calcium ions. Following their theory, the higher affinity of local anaesthetics for these binding positions prevents the structural alterations in the lipoproteins which are necessary for sodium influx. Nuhn [45] also states that the displacement of the membrane-bound calcium ions seems to be the reaction which determines the effect of the local anaesthetic. According to Seeman [50], the displacement of calcium ions by local anaesthetic also leads to measurable membrane expansion and an increase in surface tension, both of which correlate directly with the local anaesthetic effectiveness of a substance [10, 54].

Through their penetration into the hydrophobic areas of the membrane, local anaesthetics also lower the phase-transition temperature of lipids between gel phase and fluid phase [34, 45, 50]. This leads to an increase in the fluidity of the lipids surrounding the sodium channels and to expansion of the membrane. Lee [34] considers this fluidization to be the true pharmacological mechanism of local anaesthetics. Finally, the membrane expansion itself is discussed as at least an essential component of the mechanism [45, 59]. Local anaesthetics have this effect in common with general anaesthetics and some neuroleptics. With volatile anaesthetics, both membrane expansion and anaesthetic effect are antagonized by hyperbaric pressures. The lack of proof of an analogous effect of hyperbaric pressures on local anaesthesia has in the past been raised as an important argument against such a mechanism of effectiveness being common to all anaesthetics, but Kendig and Cohen [33] have shown that this pressure antagonism is in principle also effective in local anaesthesia.

Table 1a. Classification of nerve fibres (Gasser and Grundfest)

Type	Function	Diameter	Conduction speed (m/s)
Aα	Voluntary motoricity Muscle spindle afferent	15 μm	70–120
Aβ	Skin afferent for touch and pressure	8 μm	30– 70
Aγ	Muscle spindle efferent	5 μm	15– 30
Aδ	Skin afferent for temperature and pain	3 μm	12– 30
B	Sympathetic preganglionic	3 μm	3– 15
C	Skin afferent for pain Sympathetic postganglionic	1 μm (not myelinated)	1

Table 1b. Classification of sensory nerve fibres (Lloyd-Hunt)

Group	Function	Conduction speed (m/s)
I	Primary muscle spindle afferent and tendon afferent	70–120
II	Skin mechanoreceptors	25– 70
III	Deep pressure sensitivity of muscle	10– 25
IV	Non-myelinated pain afferent fibres	1

Classification of Nerve Fibres by Conduction Velocity and Function

The fibres of the peripheral and autonomic nervous systems were classified by speed of conduction and assigned to particular physiological functions by Gasser and Grundfest [25] (Table 1a). The classification of sensory nerve fibres according to Lloyd-Hunt [36] is also commonly used (Table 1b).

Latency and Regression Times of Regional Anaesthesia

Local anaesthetics usually come in the form of aqueous solutions – hydrochlorides, phosphates or carbonates of only sparingly water-soluble (weak-base) substances. The dissociation base : cation ratio is substance-specific and pH-dependent for every local anaesthetic. The pH value of the aqueous solutions is below the physiological value. In regional anaesthesia of a peripheral nerve the water-soluble salt diffuses in the interstitial fluid through the connective tissue sheaths of the nerve and reaches the membranes of various nerve fibres, the concentration gradient and time factor depending on distance. In the physiological pH range the local anaesthetic dissociates and the lipophilic base can penetrate the nerve membrane.

According to Narahashi and Frazier [38], the blockade of the sodium channels takes place on the inner (axonal) side of the membrane through the agency of the ionized portion (cation) of the dissociated local anaesthetic. The non-ionized base portion is the "transport form" of the local anaesthetic through the membrane.

The *latency time* of local anaesthesia, defined as the time elapsing between injection and total interruption of conduction, is essentially a function of the concentration of the anaesthetic, the extraneural diffusion distance (distance from injection cannula to nerve), the intraneural diffusion distance (diameter of nerve) and the dissociation constant of the anaesthetic. The latency times for the blockade of individual fibres in the nerve are therefore determined by the topographical relationship of the fibres to the surface of the nerve (diffusion distance and concentration gradient) [12]. In addition, according to Gasser and Erlanger [24], different types of nerve fibres have different sensitivities to local anaesthetics, and this influences their latency time [12]. The sensitivity is said to be inversely proportional to axon diameter and degree of myelination of the fibres. Thus, the first fibres to be blocked by regional anaesthesia are non-myelinated pain and sympathetic fibres, followed by the weakly myelinated Aδ pain- and temperature-sensing fibres, and finally by the Aβ pressure- and touch-sensing fibres and the Aα fibres of the voluntary motor system.

The *regression* of the blockade begins with the inversion of the concentration gradient between membrane and extracellular space and ends clinically with the complete return of conduction. Regression is also said to have a temporal association with the morphology of the various fibres, conduction returning in the reverse order to that in which it was blocked, proportional to axon diameter and degree of myelination.

From this mirror-image behaviour of latency and regression times, the conclusion was reached that the differences in sensitivity among the fibres cannot relate to different surface/volume ratios [12]. If this were so, penetration by local anaesthetics of fibres with small axonal diameters would certainly be favoured, but analogously there would also be rapid return diffusion and thus a relatively early onset of regression. Nathan and Sears and Tasaki [40, 41, 60] therefore explained the differences in sensitivity by a myelination-dependent "security factor of nerve conduction".

Methods for Measuring Effectiveness of Local Anaesthesia on Groups of Fibres in Nervous System

Animal Experiments

Since the fundamental experiments by Gasser and Erlanger [24], all animal experiments employ the same basic principle of measurement: The nerve preparation is put into a measuring chamber and a short length of it is brought into contact with the local anaesthetic. One end of the nerve is electrically stimulated, at the other end the compound action potentials of the fibres are monophasically conducted to a cathode-ray oscillograph where they are amplified and imaged. Owing to the differences in conduction velocity between the fibre groups, the recording of a single stimulation of a mixed nerve registers an individual compound action potential for each group. The effect of anaesthesia is measured by the amplitude of these compound action potentials. The important variables in this procedure are choice and preparation of the nerve, method of stimulation, length of nerve contacted and concentration of local anaesthetic [11, 16, 18].

Clinical Experiments

There are standardized procedures of measurement for the clinical testing of effectiveness of local anaesthesia. Voluntary motor function is assessed by muscle strength, EMG activity or conduction velocity of motor fibres (also by means of EMG) [22, 42–44, 62]. These methods give almost equal results [42], and allow precise quantification of the course of blockade of the Aα fibres. The blockade of the sensations of cold, heat, pain and touch can only be tested qualitatively, i.e. through determination of latency and regression times. Standardized methods of stimulation [22] and cooperative subjects are necessary for this, and therefore such determinations are considered relatively unreliable [53]. The blockade of sympathetic fibres is assessed indirectly in clinical experiments, no sympathetic function being susceptible to direct measurement. Sympathetically regulated secretion of sweat is tested with indicators (bromocresol green or ninhydrin); when no more indicator stains the skin the sympathetic system is considered to be blocked [7, 42]. This method is time-consuming and imprecise [42]. Most procedures for registering total sympathetic blockade therefore employ the increase in skin blood perfusion which results from vasomotor blockade. In the hands and feet, vasomotor tone is governed completely by sympathetic innervation [49]. Every vasodilatation in these regions corresponds to a decrease of impulse frequency in the sympathetic fibres [52]. This impulse frequency in turn correlates quantitatively with the liberation and elimination of the transmitter substance noradrenaline [20]. Total sympathetic denervation therefore results in interruption of neural regulation of vasomotor tone [35]. The loss of tone in the precapillary resistance vessels causes an increase in pulse wave amplitude of the skin capillaries [28], measurable with photo-electric cells, and in skin temperature, measured using a skin electrothermometer [22, 42, 62], or without touching the skin by means of telethermography [57]. These indirect methods were mostly employed for determination of sympathetic latency and regression times, as no defined starting temperatures or pulse wave amplitudes were established for measurement of temperature changes in the course of blockade. According to Nolte [42], the sympathetic system is blocked when the skin temperature rises to 1.5 °C above that in non-blocked areas; according to Fruhstorfer [22], it can be considered blocked when the skin temperature has reached 25% of the measured end temperature. These definitions are arbitrary, and it is astounding that the latency and regression times established in this way are used for comparison with those of other functions.

Theory of Differential Block

Experimental and Clinical Foundations

The theory of the differential sensitivity of fibres in the peripheral nervous system essentially goes back to the experimental investigations that Gasser and Erlanger published in 1929 [24]. They proved that when isolated peripheral nerves are stimulated electrically under the influence of increasing concentrations of cocaine, the compound action potentials of slow-conducting fibres dissappear before the compound action potentials of the fast-conducting groups of fibres show any detectable change.

These findings were essentially also confirmed with local anaesthetics of the ester and acid amide types [31, 37, 39, 48, 61]. The blockade of slow-conducting fibres by low concentrations of local anaesthetic while fast-conducting fibres continue to function normally has since been termed "differential block". The experimental investigations of differential block with the aid of cathode-ray oscillography of the action potentials of isolated nerves were confirmed by clinical observations which reach back to the historical beginnings of regional anaesthesia [4, 8, 16].

In vivo experiments with peripheral, spinal and extradural regional anaesthesia, in which comparisons of the latency and regression times of the blockade of the various sensations, vasomotoricity and voluntary motoricity were made, also confirmed the same differential sensitivity of the fibres responsible for these functions [7, 22, 23, 27, 29, 42, 44, 47, 62].

The action of regional anaesthesia begins subjectively with a feeling of warmth and objectively with a corresponding rise in skin temperature. Then successively come clear decreases in the sensations of cold and of heat and cutaneous pain, the loss of the sensations of pressure and touch, and finally the loss of voluntary motoricity. The regression of the blockade takes the reverse order. Because of the unreliable investigation methods, the sequence of blockade and regression of the sensations of cold, heat and pain has been variously described [53]. Nevertheless, comparisons of latency times for the blockade of vasomotoricity, pain, pressure, and voluntary motoricity correspond to the increasing axonal diameters and degrees of myelination of the nerve pathways conducting these functions.

Spinal and extradural anaesthesia displays a special variant of differential block. The dilution of the local anaesthetic in the cerebrospinal fluid in spinal anaesthesia and the great diffusion distances in the epidural space in epidural anaesthesia result in steep concentration gradients in the spinal segments, increasing with distance from the lumbar injection site. Spinal segments closest to the puncture show not only the shortest latency time, but also the strongest blockade and the greatest duration of action. It is here that motoricity and the sensation of pressure are most often blocked; in more remote segments, motoricity and the sensation of touch are conserved although analgesia is total. After the last dermatome of cutaneous analgesia come several more with isolated blockade of the sensation of temperature and increased cutaneous blood perfusion.

Contradictory Findings

The theory of differential sensitivity of nerve fibres to local anaesthetics is not unchallenged. Gasser and Erlanger [24], Toman [61] and Heinbecker et al. [31] were not able to confirm the validity of the theory in all experiments. In their in vitro investigations, Everett et al. [18, 19] varied the intensity and duration of the stimulation of the nerves and found, with otherwise comparable methods, than non-myelinated C fibres were even less sensitive than A fibres. Douglas and Ritchie [17] were the first to express profound doubt concerning the methodology of previous in vitro investigations, claiming that quantitative comparison of the compound action potentials in a peripheral nerve containing fibres of various conduction velocities is not possible. The oscillations in potential of individual fibre groups − so runs their argument − must, on the basis of the differing conduction speeds and the differing initial values of the compound action potentials, be governed by differing temporal dispersion. Franz and Perry [21] compared the compound action potentials of isolated nerve fibre groups and varied the length of fibre brought into contact with the local anaesthetic. It turned out that a concentration-dependent differential block could be proved only when the contact length was limited to 2 mm. When it was increased to over 4 mm, the blockade occured simultaneously in all fibre groups, independent of the concentration of local anaesthetic.

Another factor speaking against the existence of differential sensitivity dependent on axonal diameter and degree of myelination was found by Heavner and de Jong [30]. They proved that the myelinated preganglionic B fibres of the sympathetic nervous system are blocked by significantly lower concentrations of local anaesthetic than are non-myelinated postganglionic C fibres. These findings were recently confirmed by Rosenberg et al. [48], who investigated in the rabbit the influence of increasing concentrations of local anaesthetic on the action potentials of B and C fibres in the sympathetic trunk and motor A fibres in the phrenic nerve. The comparison of postganglionic C fibres with motor A fibres confirmed, however, the greater sensitivity of the C fibres to bupivacaine and chloroprocaine. Finally, Gissen et al. [26], using almost exactly the same methods, determined a completely new sequence of differential sensitivity in nerve fibres. In their experiments, the greatest resistance to the local anaesthetics bupivacaine, etidocaine, lidocaine and tetracaine and the biotoxins tetrodotoxin and saxotonin was shown by non-myelinated C fibres from the rabbit vagus nerve; less resistance was displayed by preganglionic B fibres from the same nerve, and the most sensitive fibres were fast-conducting A fibres from the sciatic nerve. They left no doubt of the invalidity of the formerly held theory of differential block [26]. Doubt was expressed, however, by de Jong, who commented on their work in the same journal. His concern related to methodology – the in vitro investigations were carried out not at 37 °C, but at room temperature – and the discrepancy of the results from clinical experience. He ended by saying: "There remains plenty to be done yet before the book on differential nerve block can be closed." [13].

Doubt was also recently cast on the validity of the theory by the results of clinical investigations [57]. The vasomotor tone of the skin vessels was recorded indirectly by telethermography during the latency period of extradural and peripheral regional anaesthesia. It emerged that the radiation of heat by the skin increases until the blockade of the voluntary motor system reaches its maximum. When motor blockade was more effective on one side than the other at the end of the latency period in peridural anaesthesia, thermography showed corresponding differences in cutaneous blood perfusion, although total analgesia was registered in the skin on both sides. If the anaesthesia was then intensified with a further injection, the difference in motor blockade was abolished by an increase on the formerly less blocked side, and thermography showed dissappearance of the discrepancy in blood perfusion. A further indication of identical sensitivity of motor A fibres and sympathetic C fibres to local anaesthetics was supplied by thermographic documentation of sympathetic blockades in the upper extremity [57]. In patients with Raynaud's disease, sympathetic innervation of one hand was interrupted by stellate block, that of the other hand by axillary plexus anaesthesia. Even with total analgesia, the hand under plexus anaesthesia, which involved low concentrations of local anaesthetic, displayed a lesser degree of sympathetic block than the hand under stellate block. Only with a stronger motor blockade did the intensity of the sympathetic block under plexus anaesthesia approach that under stellate block. All these half-quantitative, thermographic observations disagreed with the results of previous clinical-experimental investigations.

Statement of Problem

As obviously any sequence of differential sensitivity can be established in in vitro investigations using cathode-ray oscillography of the compound action potentials, conclusive proof of the differential block seems possible only in vivo. So far, however, in vivo investigations have used disparate methods to compare blockade and regression times of nerve functions. This is seen most clearly in the comparisons of the effect of anaesthesia on the function of the voluntary motor system and the vasomotor system. Voluntary motoricity was measured as a defined performance, but the time which elapsed until it was completely blocked was compared with a vasomotor latency time which was measured by an arbitrarily chosen rise in temperature [22, 42]. For the tested latency times of perception of sensations, it is not even certain that the loss of perception is equated with interruption of nerve conduction.

The question of whether differential sensitivity of nerve fibres exists is therefore only to be answered, on the foundation of well-known methods of measurement, by quantitative comparison of voluntary motor and vasomotor blockade. The latter, however, can be comparably quantified only if we succeed, through maximal stimulation of vasomotor tone, in achieving measurement of a loss of performance equivalent to that in the voluntary motor system. Moreover, the measurement of vasomotor performance must take place in a representative patch of skin on the hands or the feet, as it is only there that the tone of the vascular smooth muscle is governed by sympathetic innervation alone. Measurement over large areas is necessary, as only in this way can a complete picture of the blockade of all fibres be given, just as the measurement of muscle power in increasing and decreasing blockade gives only the total value for the fibres as a whole, although the existence of concentration gradients means that different sections are blocked to different degrees. We therefore investigated latency time and regression in peripheral regional anaesthesia under experimental conditions which guarantee approximately equal conditions for measurement of voluntary motoricity and sympathetically innervated vasomotoricity.

Methods

Subjects and Parameters

Thirteen healthy men and women aged between 23 and 37 years volunteered for peripheral regional anaesthesia. In six subjects the latency time was studied from the point of view of optimizing methods. In the other seven, the blockade and regression of vasomotoricity and voluntary motoricity were quantitatively measured and latency and regression times precisely defined. The latency and regression times of the sensations of cold, pain and touch were recorded on the basis of subjective statements.

Techniques of Anaesthesia

Peripheral nerves were chosen in which continuous perfusion and simple washing out of the local anaesthetic are possible. A further condition was a terminal autonomous innervation zone in the region of the hands or the feet. The continuous blockades of the nerves were carried out using special double cannulae designed especially for this experiment. The inner cannula has an external diameter of 0.55 mm and consists of a Sprotte atraumatic spinal cannula[1] shortened by 4 cm. Its opening is on the side of the shaft just before the conical point. the outer cannula, a blunt metal sleeve with an external diameter of 0.9 mm, ends shortly before the side opening of the inner cannula. After atraumatic puncture with this combination needle, the pointed inner cannula was removed and the blunt metal sleeve advanced in the perineural connective tissue parallel to the course of the nerve to avoid accidental dislocation of the cannula during movement of the extremities.

Blockade of Median Nerve

After local anaesthesia of the skin, the brachial artery was palpated in the distal sulcus of the biceps muscle a few centimetres above the lacertus fibrosus. The double cannula was introduced medially and tangentially to the brachial artery into the fascial interstitium common to the artery and the median nerve. The inner cannula was removed and the remaining blunt outer cannula fixed to the skin and connected via a perfusion lead to an electromechanical injection pump. In this way the arm could be moved freely without any risk of injury to the nerve.

---

1 Manufactured by Gebrüder Pajunk, Geisingen, West Germany and distributed by Hell & Co., Nuremberg, West Germany

Blockade of Brachial Plexus

The neurovascular sheath of the axillary brachial plexus was cannulated and connected with the infusion pump in the same way. The puncture and the fixation of the cannula were far enough away from the axilla for the arm to be freely mobile.

Blockade of Tibial Nerve and Common Peroneal Nerve

Puncture was carried out with the double cannula on the medial edge of the biceps femoris muscle, in the upper third of the triangle formed above the popliteal space by the medial edge of the semimembranous muscle and the biceps femoris muscle. After puncture of the popliteal fascia in a steeply cranial direction, paraesthesia of the tibial nerve or common peroneal nerve was produced, the inner cannula removed and the remaining sleeve advanced further cranially. After fixation of the sleeve and connection to the injection pump, the knee too could be moved freely. This technique was developed specially for this investigation. The injection of 30–40 ml local anaesthetic solution via this route leads to anaesthesia of all distal branches of the sciatic nerve. With the exception of the medial region of the ankle, the foot is completely anaesthetized. On the dorsum of the leg, anaesthesia extends proximally to the popliteal space after injection of smaller volumes of local anaesthetic (20–30 ml), sometimes as far as the skin fold over the lower edge of the gluteus maximus muscle after larger volumes (30–40 ml).

Dosage and Choice of Local Anaesthetics

All nerve blockades were carried out using mepivacaine without added vasoconstrictor. Mepivacaine is a standard local anaesthetic with a mean duration of effect of 120 min for a 1% solution and 160 min for a 2% solution [47]. In contrast to other local anaesthetics, in the usual concentrations (0.5%–2%) it has no significant dilating effect of its own on vascular smooth muscle [1, 14, 32]. Perineural infusion was carried out with 0.5%, 1% or 2% mepivacaine. No blockade could be achieved with the 0.5% solution. However, in order to be able to register any possible concentration dependence of the differential block, the perineural connective tissue was perfused slowly with the local anaesthetic so that diffusion to the individual fibres in the nerve was considerably delayed and displayed clear concentration gradients.

In order to establish the optimal speed of perfusion, six preliminary investigations were carried out in which the process was followed only until the end of the latency period. Depending on the position of the cannula in relation to the nerve, doses of between 0.5 and 1 ml/min proved sufficient on the one hand to overcome the simultaneous resorption of the local anaesthetic and achieve a complete blockade, and on the other hand to provide a protracted latency period with the greatest possible concentration gradients and an adequate span of time for the measurements. In three of the seven subjects in whom latency and recovery were investigated, the local anaesthetic was washed out with an injection of physiological saline containing hyaluronidase when complete blockade of the nerve had been achieved; in the other four the blockade was allowed to regress spontaneously.

Measurement of Voluntary Motoricity

Quantitative determination of voluntary motoricity was carried out by measurement of muscle power, using spring balances with different ranges (0–1 kg, 0–12 kg and 0–20 kg) according to the intensity of the blockade. The measurements were made by two independent investigators. In blockade of the median nerve, the hand and the proximal phalanx of the index finger were fixed supine on a horizontal surface. The middle and distal phalanges were held in 90° flexion with a broad loop from a spring balance around the distal phalanx and the power exerted against the horizontal pull of the balance was measured. In blockade of the brachial plexus this measurement was also performed, in analogous manner, in the fifth finger, in order to test the power in a muscle representative of those supplied by the ulnar nerve (flexor digiti minimi brevis manus and flexor digitorum profundus 4 and 5). In blockade of the tibial nerve and the common peroneal nerve, the power of the foot extensors (common peroneal nerve) and the flexor hallucis longus (tibial nerve) was tested against the horizontal pull of the spring balance. Measurements of the power present before the blockade were made 30 min before and immediately before the infusion pump was turned on.

Measurement of Sympathetic Activity

Cold Stimulation of Sympathetic Activity

The main function of the sympathetic innervation of the blood vessels in the regions of the hands and feet is regulation of temperature. Heat stimuli decrease sympathetic activity, while cold can be regarded as a suitable promoter. The reflexive vasoconstriction induced by cold reaches its maximum at a skin temperature of 15 °C; at lower temperatures there is vasodilatation (cold-induced dilatation through paralysis of the vascular smooth muscle) [28, 55].

The subjects immersed their hands or feet intermittently for 2 min in ice-water, then dried them on terry towels. After a further 2 min, radiation of heat was measured by telethermography. The intermittend cooling was continued until several consecutive thermographies showed a constantly low level of heat radiation before the investigation was started. According to the subject's sensitivity to cold, cooling the hands or feet down to a constant minimal temperature took between 20 and 40 min. The cooling was continued on the same pattern while the investigations were under way.

Measurement of Vasomotor Tone by Telethermography

The slackening of vascular muscle tone under nerve blockade leads to increasing perfusion of the skin capillaries and thus to a proportional increase in heat radiation in the area innervated by the nerve in question. Like an object coloured black, human skin absorbs electromagnetic waves of from 3 to 15 μm wavelength, and is consequently also an ideal radiator of heat [2, 3]. Thus, by means of thermography, changes in blood perfusion of the skin can be measured indirectly, through heat radiation, without contact, and through active cooling without potentially disturbing changes in environmental temperature.

Technical Details of Measuring Apparatus

The Philips thermograph that we employed consists of a camera, a display system, and colour and black-and-white monitors. The liquid-nitrogen-filled camera uses single elements of indium antimonite to register electromagnetic heat radiation with a thermal resolution of less than 0.08 °C and a focussing range of 20 cm to infinity. The heat image, made up of lines, is projected through the display system onto the TV monitors with a frequency of one image per second. The working temperature lies in the range 15 °–40 °C. On the colour monitor, the heat image is broken up into ten colour levels, each colour designating areas of similar temperature. The "width" of the temperature bands can be set anywhere between 0.2 ° and 1.4 °C. Any temperature in the whole working range can be selected as ground temperature (black level) and appears on the screen as absolute temperature value in degrees Celsius. With the aid of a colour scale at the edge of the image, similarly coloured areas of the thermogram can be identified, and the temperatures they represent can be calculated from the black-level temperature and the width of the temperature bands.

Skin Areas Investigated

The temperature measurements were made over the skin of the sole of the foot or the palm of the hand. The anaesthetized nerves govern large areas of autonomic innervation in these regions: the whole of the sole is supplied by the tibial nerve, the innervation of the volar surface of hand and fingers is shared by the median and ulnar nerves. At the borders with areas supplied by neighbouring nerves, there are narrow zones of mixed sensory and autonomic skin innervation. Figure 1 shows thermograms, one black-and-white and one colour, of two total nerve blockades, including innervation boundaries with neighbouring nerves.

Quantitative Evaluation of Thermograms

When a constant minimal temperature of the skin had been reached through cooling, the working range of the camera was fixed accordingly. The black level was so set that at the lowest colour level of the thermogram only one part of the area under investigation became visible. Only with this setting could the maximum measurement range of 14 °C be used without altering the black-level temperature; however, in a few investigations the black-level temperature still had to be raised a few degrees Celsius during the measurements. With constant black-level temperature, the temperature development could be followed visually and roughly calculated by comparing the isotherms of consecutive thermograms. In addition, the display system offers two different methods of establishing precise values for graphic documentation of the changes in temperature:

1. Rectangles of any desired size can be called up in any region of interest on the thermogram. The mean temperature of these areas appears as differential temperature to the black level under the designation of "Index" at the bottom edge of the monitor image. For our measurements, we used rectangles of constant size and position over representative areas of the autonomic innervation zone. For example, the series of thermograms in the Investigation 11 (Figs. 15, 16) were evaluated in this way, and the values determined

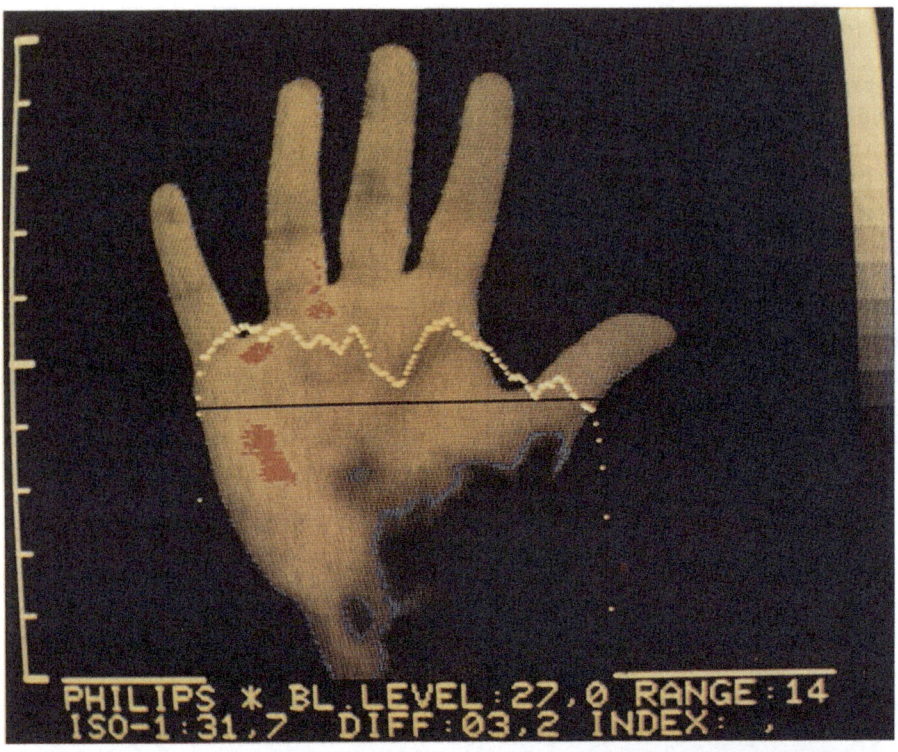

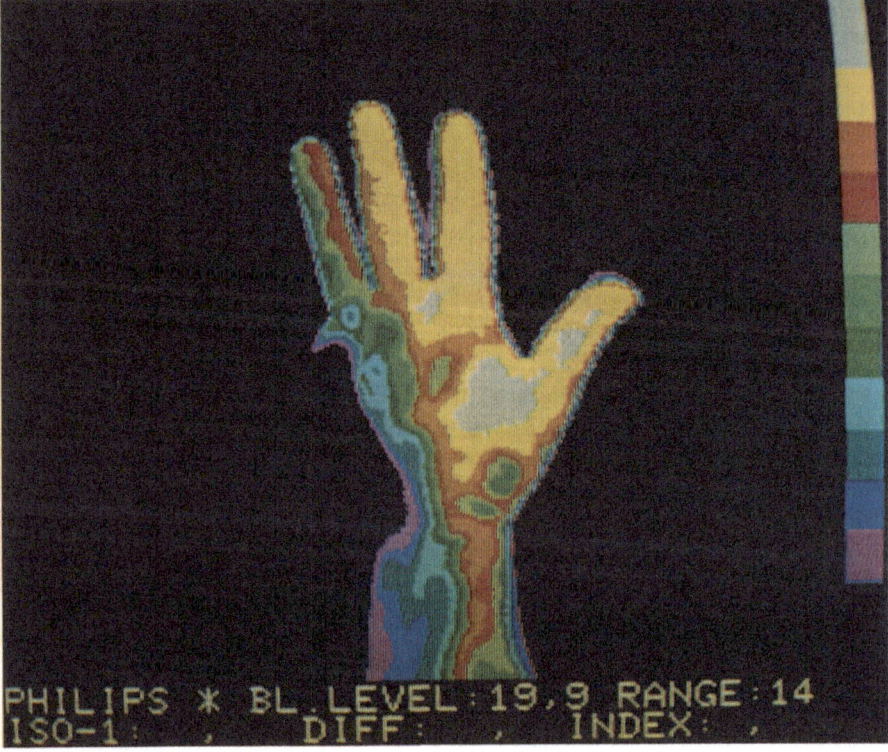

Fig. 1.

were used to produce the graphic presentation of temperature course in Fig. 14. In the black-and-white thermograms of this series, the index values of a representative area over the autonomic supply zone of the ulnar nerve are documented. The corresponding mean values from the zone of the palm supplied by the median nerve were taken from simultaneous colour thermograms. After completion of pre-cooling of the hand, the mean temperature (black level plus index value) in this test is 17.06 °C in the region of the ulnar nerve (thermogram 1 of the black-and-white series). It rises after perfusion of the nerve plexus to a maximum of 30.38 °C (thermogram 9), and then sinks again after regression of the blockade to a constant 16.98 °C, similar to the initial value (thermogram 18). After rounding these values up and down (17 °C–30.5 °C), the range of temperature change was calculated at 13.5 °C.

2. In the black-and-white monitor image, two isotherms of 0.1 °C in width can be superimposed. An example is shown in Fig. 1. An orange-coloured isotherm is used to mark the maximum temperatures in the thermogram, which are then used to assess the course of the blockade. The values recorded appear as differential temperature to a second, blue-coloured isotherm (Iso-1), which is set so that it describes the perimeter of the thermogram. The calculation of range of maximum temperature was analogous to that used in the first method.

Graphic Presentation of Measured Values

For simultaneous graphic presentation of the effect of anaesthesia on voluntary motoricity (Aα fibres) and vasomotoricity (C fibres), all values of power and temperature were calculated as percentages of the initial values. For voluntary motoricity this was the maximum power of the reference muscle, for vasomotoricity the temperature at the bottom end of the established range.

◀ **Fig. 1.** (*upper*): Example of a black-and-white thermogram. Complete plexus brachialis anaesthesia in the terminal innervation areas of the ulnar and median nerves: no voluntary motion, no sense of touch, maximal radiation of heat (vasomotor blockade). The *blue isotherm* in the region of the proximal thenar and the radial forearm marks the boundary with the innervation area of the lateral cutaneous nerve of the forearm, which because of the greater extraneural diffusion distance in the axillary plexus anaesthesia is less intensively blocked. Despite the absence of the sensations of pain and cold, a clear vasomotor reflex is visible here, and the sensation of touch is also retained in this region. The base temperature (*black level*) is 27 °C, the absolute temperature of the blue isotherm is 31.7 °C and the maximal temperature of the thermogram (*orange isotherm*) is 31.7 °C + 3.2 °C = 34.9 °C.

(*lower*): Example of a colour thermogram. Complete median block with autonomic area (*yellow and white*), mixed innervation with fibres from the ulnar nerve (*left*) and the lateral cutaneous nerve of the forearm (*proximal thenar and radial forearm*). The area of autonomic innervation of the non-anaesthetized ulnar nerve is 19 °C cooler than that of the anaesthetized median nerve and therefore outside the range of measurement of the camera (*fifth finger and hypothenar*). The colour scale on the *right edge* of the image shows isotherms of increasing temperature from bottom to top. The base temperature (*black level*) is 19.9 °C, the range of measurement 14 °C. The isotherms of the ten-level scale are therefore 1.4 °C wide.

Definition of Latency and Regression Times

The *latency time* of the motor blockade was defined as the time between the commencement of perineural perfusion and the complete cessation of voluntary motion in the corresponding reference muscle, the *regression time* as the period between the first voluntary motion and the regaining of the initial power. The last weak movements before blockade became complete and the first after the onset of regression could not be quantified by measurements of power, but could be clearly defined in terms of time.

The *latency time* of the vasomotor blockade was defined as the period between the commencement of perineural perfusion and the point where the thermographically measured temperature ceased to rise. Analogously, *regression* began with the recognizable decrease in heat radiation and ended with the recording of constant minimum temperatures essentially comperable with the initial temperatures. Precise definition of these times was achieved predominantly by optical comparison of consecutive colour thermograms. Representative mean and maximum temperatures did not allow reliable determination of these time points in all investigations, as the last and first temperature changes take place on the periphery of the regions supplied by the nerves in question. The determination of latency and regression times of vasomotor blockade by comparison of colour thermograms is described using the examples of median blockade (Investigation 7) and plexus anaesthesia (Investigation 9).

The definition of the blockade times for the sensations of cold, pain and touch is not based on precise measurements; the loss of these sensations under the nerve blockade is superimposed by new qualities of sensation which do not fit exactly into this category (paraesthesias). For instance, at one point during the blockade the ice-cold cooling water is no longer perceived as cold, only as wet; it is practically impossible to separate these sensations for the purpose of determining this point. The sensation of cold was considered to be blocked when the hand or foot could be immersed in ice-water without the subject feeling any cold over the area supplied by the nerve under investigation.

Also difficult is the definition of the latency time for the sensation of pain, as strong stimuli cannot be applied without injury to the subject. The sensation of pain was tested by pinching the skin with a blunt clip, choosing areas not included in thermographic measurements in order to avoid hyperaemic effects. Blockade was considered complete when the pinching was no longer perceived as painful.

The sensation of touch was tested by lightly stroking the subject's skin while he had his eyes closed. If the subject failed to notice the stroking several times in a row, touch was regarded as blocked. The latency and regression times were entered into the diagrams as colour bands of decreasing (latency) or increasing (regression) colour density, according to the progressive nature of the loss of perception, in order to be able to compare them with the values recorded for voluntary motoricity and vasomotoricity.

Evaluation of Results

The 13 individual blockades studied differed in dose and diffusion distance, and statistical comparison of the latency times is therefore not meaningful. This investigation was carried out to test the validity of the hypothesis that there is a consistent relationship between the morphology of nerve fibres and their sensitivity to local anaesthetics. Since such a consistent relationship permits of no true exceptions, each individual investigation must be regarded as a self-contained experiment. If significant differences emerge from the individual experiments in regard to the hypothesis, this must be due to the methods employed and cannot be accorded any statistical value.

Results

Documentation

The complete results of all the investigations are shown in Figs. 2–8, 10, 11, 13, 14, 17, 18. The original measurements of changes in temperature during Investigation 11 are documented in Figs. 15 and 16. Also shown are all the colour thermograms from Investigations 7 and 9. The following date can be extracted from Figs. 2–8, 10, 11, 13, 14, 17, 18:

1. On the ordinate, in the *red* curve, *temperature* is expressed in percentage of the initial values.
2. Also on the ordinate, in the *black* curve, *power* is expressed in percentage of the initial values.
3. On the abscissa is shown the time course; the series of thermograms from Investigations 7, 9 and 11 (Figs. 10, 15, 18, 19) are numbered consecutively on the abscissa in the corresponding diagrams (Figs. 9, 14, 17).
4. The latency and regression times of the sensations of cold (K), pain (S) and touch (B) are shown as bands of decreasing and increasing colour densitiy respectively, starting at the top of the diagrams and going downwards.
5. The duration of perfusion is indicated by a double arrow under the abscissa, with dose (ml) and concentration of solution (%).

In the graphic representation of plexus anaesthesias, the courses of the blockade in the ulnar nerve and the median nerve were displayed separately.

Individual Investigations

Investigation 1

Median blockade (latency)
Subject: 33-year-old man
Dose: 1) 15 ml (2%) at 1 ml/min; 2) 5 ml (2%) at 1 ml/min

Diagram (Fig. 2). Voluntary motoricity is 40% blocked before any signs of vasomotor blockade can be recognized. The latency times of voluntary motoricity and vasomotoricity and the sensation of touch end simultaneously 66 min after commencement of perfusion, those of the sensations of cold and pain 25 min earlier.

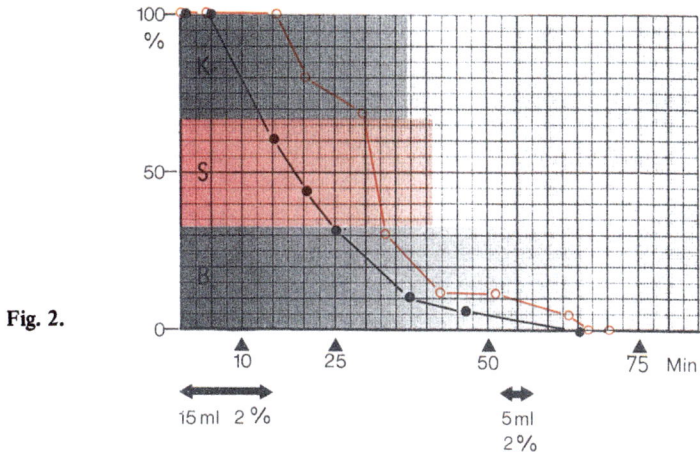

Fig. 2.

Investigation 2

    Median blockade (latency)
    Subject: 28-year-old man
    Dose: 1) 20 ml (2%) at 1 ml/min; 2) 5 ml (2%) at 1 ml/min

Diagram (Fig. 3). The blockade of voluntary motoricity and vasomotoricity begins simultaneously, and progresses almost parallel until complete interruption of conduction is achieved after 65 and 70 min respectively. The sensations of cold and pain are blocked after 55 and 58 min respectively.

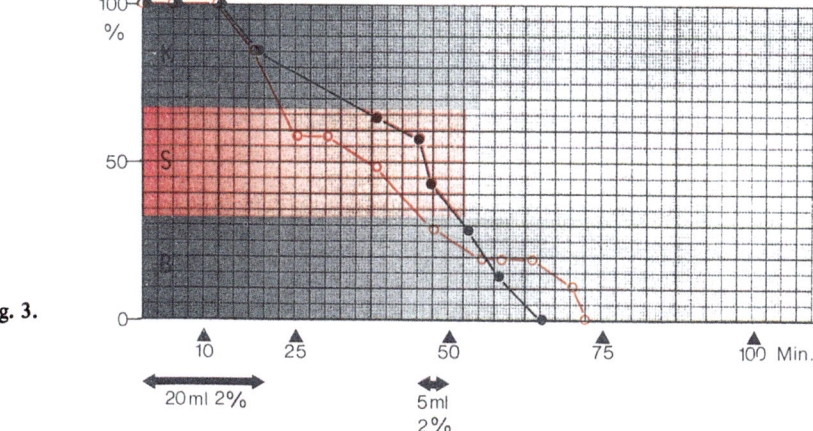

Fig. 3.

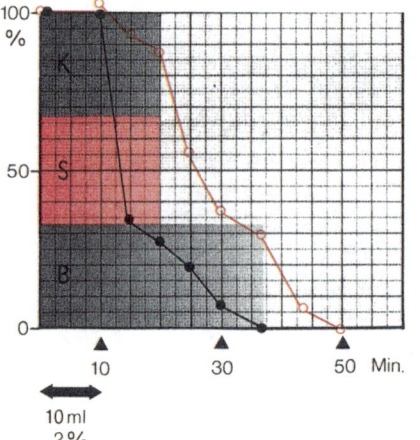

Fig. 4.

Investigation 3

Median blockade (latency)
Subject: 30-year-old man
Dose: 10 ml (2%) at 1 ml/min

Diagram (Fig. 4). Voluntary motoricity and vasomotoricity react simultaneously to the local anaesthetic, and the curves are approximately parallel, but the voluntary motor latency time is markedly shorter than that of vasomotoricity, 37 min as compared with 50 min. The sensations of cold and pain are blocked at about the same time (20 min), that of touch simultaneously with voluntary motoricity.

Investigation 4

Median blockade (latency)
Subject: 30-year-old man
Dose: 1) 10 ml (0.5%) at 1 ml/min, no effect; 2) 10 ml (1%) at 1 ml/min; 3) 5 ml (2%) at 1 ml/min

Diagram (Fig. 5). Voluntary motor and vasomotor blockade began simultaneously and the curves were parallel and almost identical until the 65th min, when after the last dose there was a slight regression of vasomotor blockade. Voluntary motoricity and the sensation of touch were blocked simultaneously after 80 min, followed shortly afterwards by vasomotoricity at 85 min. The sensations of cold and pain were blocked after only 40 min.

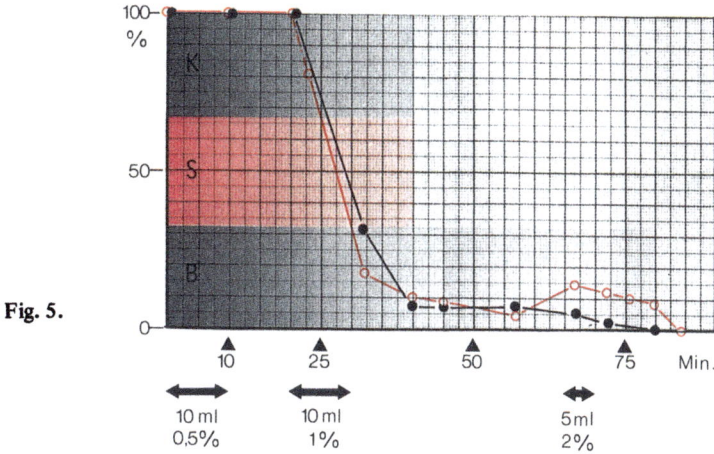

Fig. 5.

Investigation 5

   Median blockade (latency)
   Subject: 31-year-old man
   Dose: 1) 10 ml (0.5%) at 1 ml/min, no effect; 2) 10 ml (1%) at 1 ml/min

Diagram (Fig. 6). The onset of blockade is simultaneous for voluntary motoricity and vaso-motoricity, and complete blockade is reached after 52 and 48 min respectively. After complete blockade of the sensation of touch in the 42nd min, voluntary motoricity is no longer quantifiable (minimal spontaneous movements). The sensation of cold is blocked after 22 min, pain after 32 min.

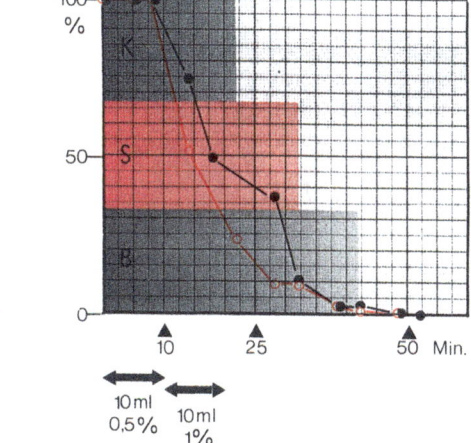

Fig. 6.

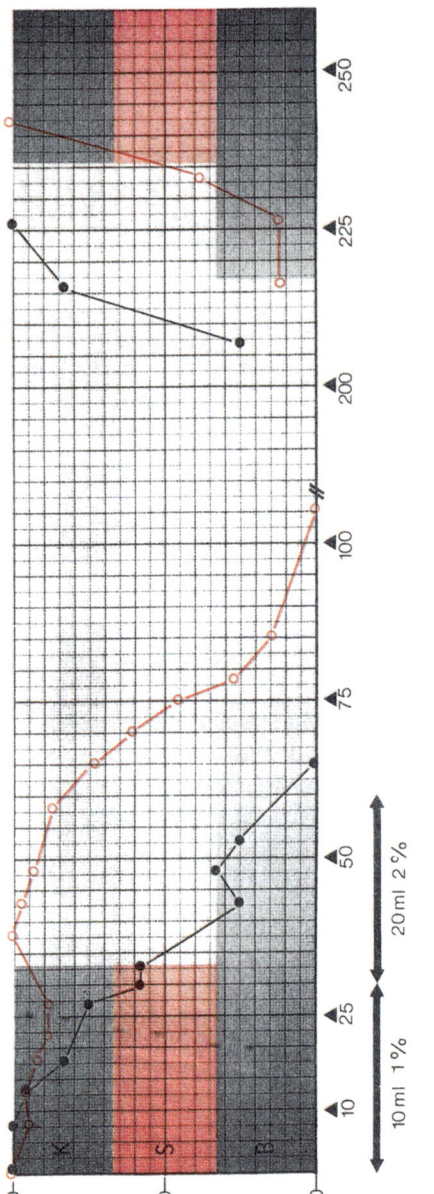

Fig. 7.

Investigation 6

Median blockade (latency and spontaneous regression)
Subject: 25-year-old man
Dose: 1) 10 ml (1%) at 0.3 ml/min; 2) 20 ml (2%) at 0.6 ml/min

Diagram (Fig. 7). Blockade of voluntary motoricity and vasomotoricity begins simultaneously. With the doubled perfusion rate, vasomotoricity recovers temporarily to the initial values,
followed somewhat later by voluntary motoricity. Complete blockade of voluntary motoricity comes after 65 min, that of the sensation of touch after 75 min and that of vasomotoricity after 105 min. The sensations of cold and pain are blocked after only 33 min. Regression
of the voluntary motor blockade begins in the 207th min. The vasomotor blockade could
not yet be measured, as spontaneous regression had begun surprisingly early. By the time intermittend cooling was resumed after a 100-min interruption, voluntary motoricity had recovered to 80% of the initial value. The regression of vasomotor blockade is therefore of only restricted usefulness in this investigation. Conduction of cold and pain stimuli begins again
in the 236th min.

Investigation 7

Median blockade (latency and regression after washing out of local anaesthetic)
Subject: 28-year-old man
Dose: 10 ml (2%) at 1 ml/min

Diagram (Fig. 8). There was simultaneous onset of local anaesthetic effect on voluntary motoricity and vasomotoricity, and simultaneous complete blockade of these functions and of
the sensation of touch. The regression of the blockade of voluntary motoricity and vasomotoricity began immediately after the local anaesthetic was washed out in the 65th min, and

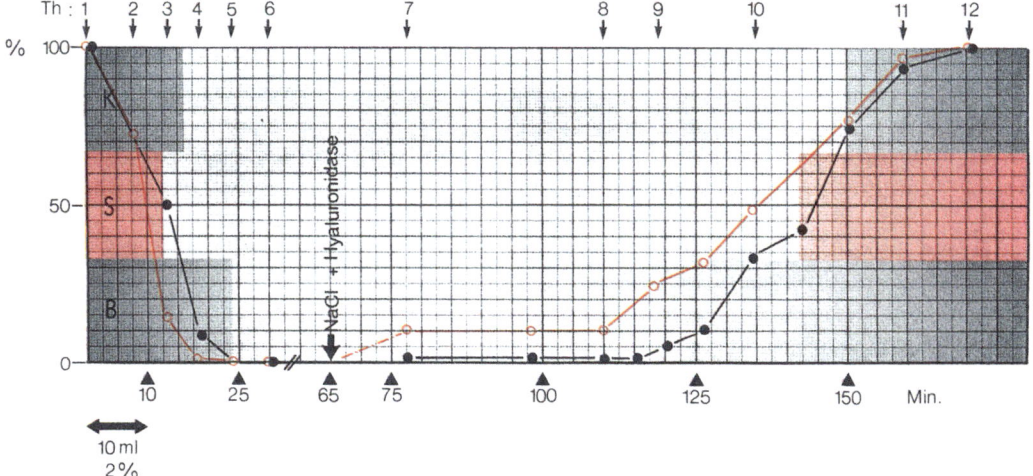

Fig. 8.

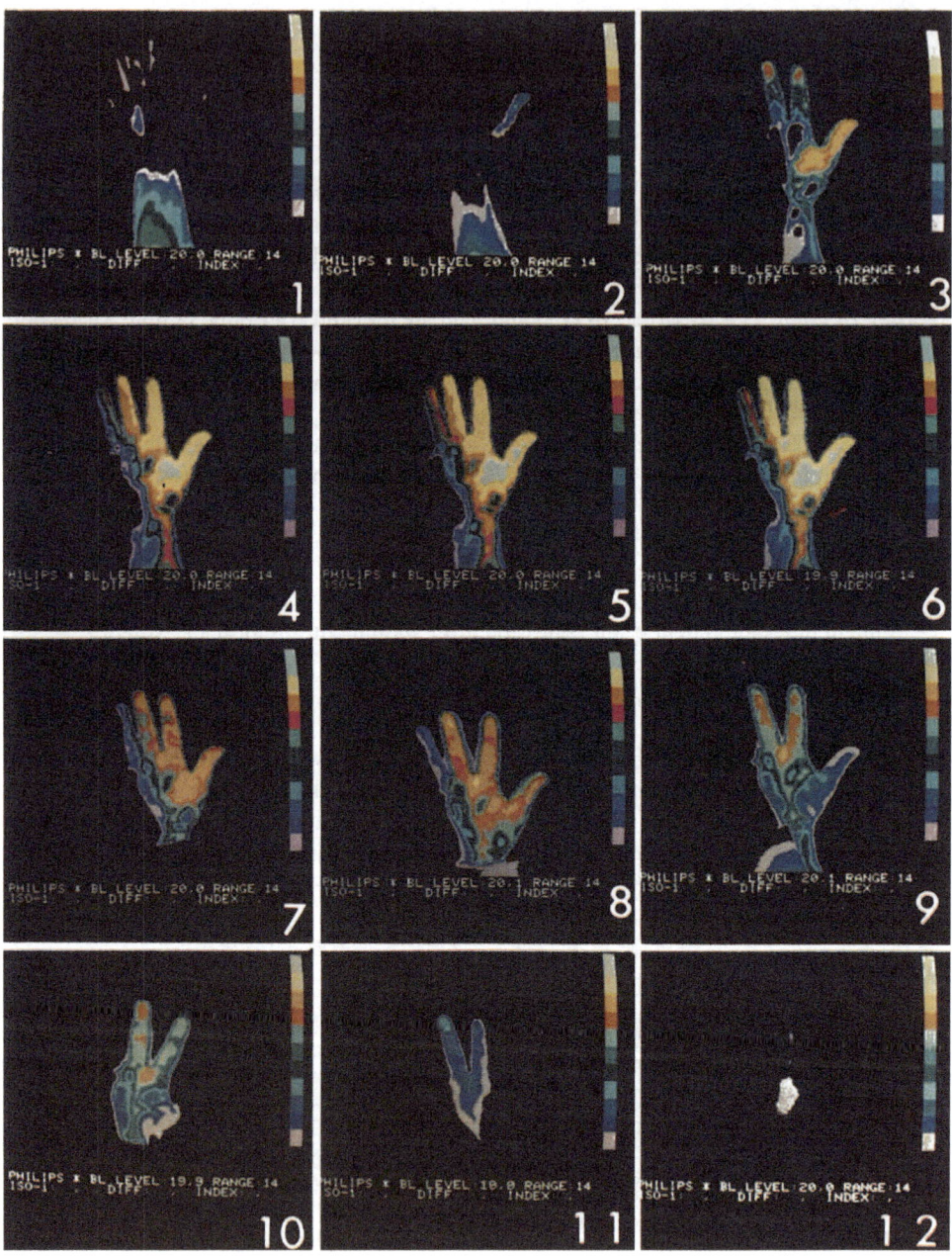

Fig. 9.

was complete (again simultaneously) after 170 min. The latency times of the sensations of cold and pain were 16 and 13 min respectively, and regression of the blockade began after 150 and 142 min respectively.

Thermograms (Fig. 9). Perfusion with the local anaesthetic begins immediately after thermogram 1. Thermograms 4 and 5 show the transition to total blockade. The temperature changes between thermograms 4 and 5 involve only the medial edge of the fourth finger, the lateral edge of the third finger and a small area over the metacarpus proximal to the base of the second finger. Thermogram 6 shows no essential differences to thermogram 5 (at 24 min), which therefore documented the end of the latency period. Thermograms 7–12 describe the course of regression of vasomotor blockade to the initial values.

Investigation 8

Brachial plexus blockade (latency)
Subject: 30-year-old man
Dose: 1) 10 ml (1%) at 1 ml/min; 2) Bolus injection of 20 ml (1.5%)

Diagram (Fig. 10). Only the course in the region supplied by the ulnar nerve is shown, as no total blockade could be achieved in the median nerve region. The effect of the anaesthesia is essentially the same on voluntary motoricity and vasomotoricity, although a minimal increase in heat radiation could be discerned for another 31 min after the total blockade of voluntary motoricity at 39 min. After bolus injection a fleeting regression of vasomotor blockade was again seen. The sensations of cold, pain and touch were blocked after 20, 26 and 34 min respectively.

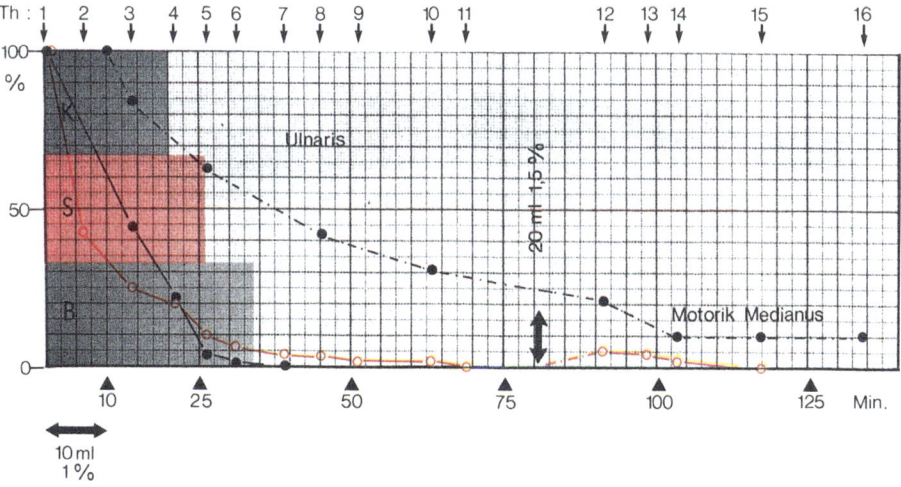

Fig. 10.

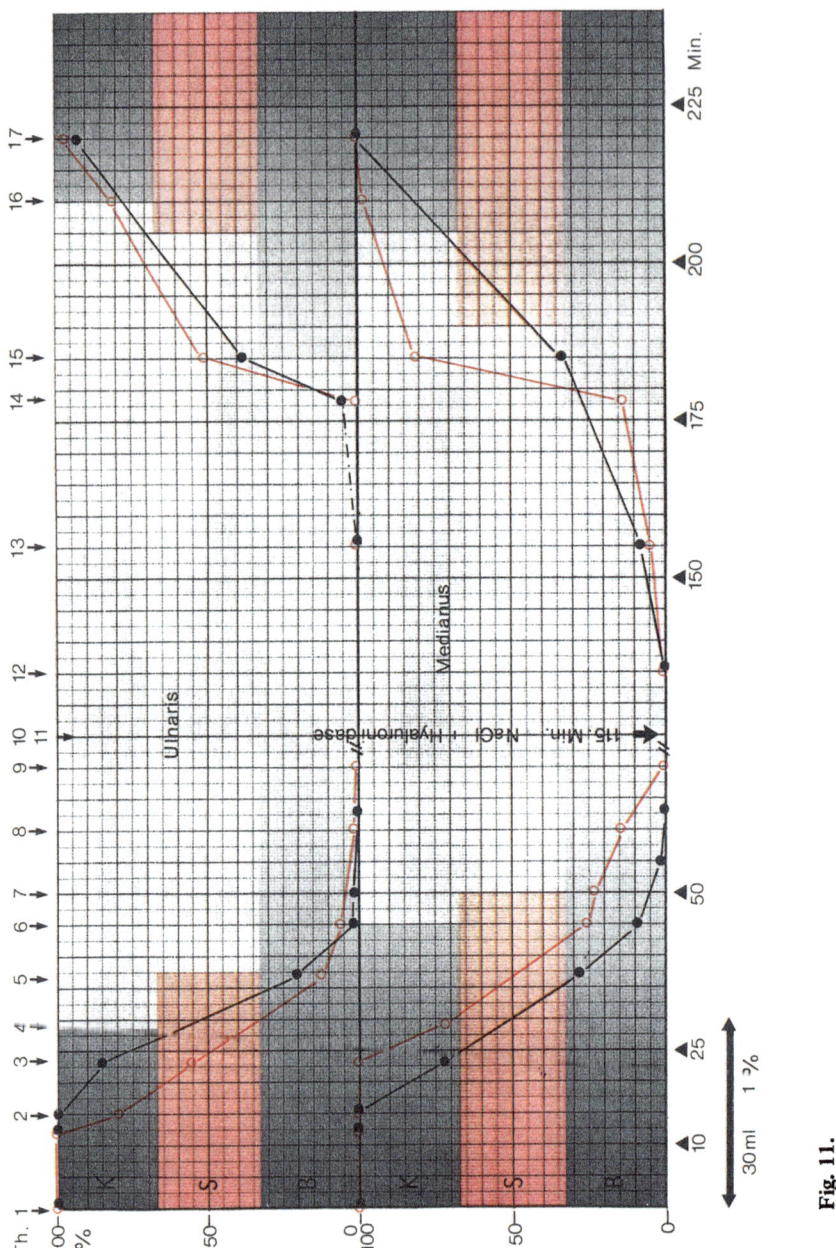

Fig. 11.

Investigation 9

Brachial plexus blockade (latency and regression after washing out of the local anaesthetic)

Subject: 31-year-old man

Dose: 30 ml (1%) at 1 ml/min

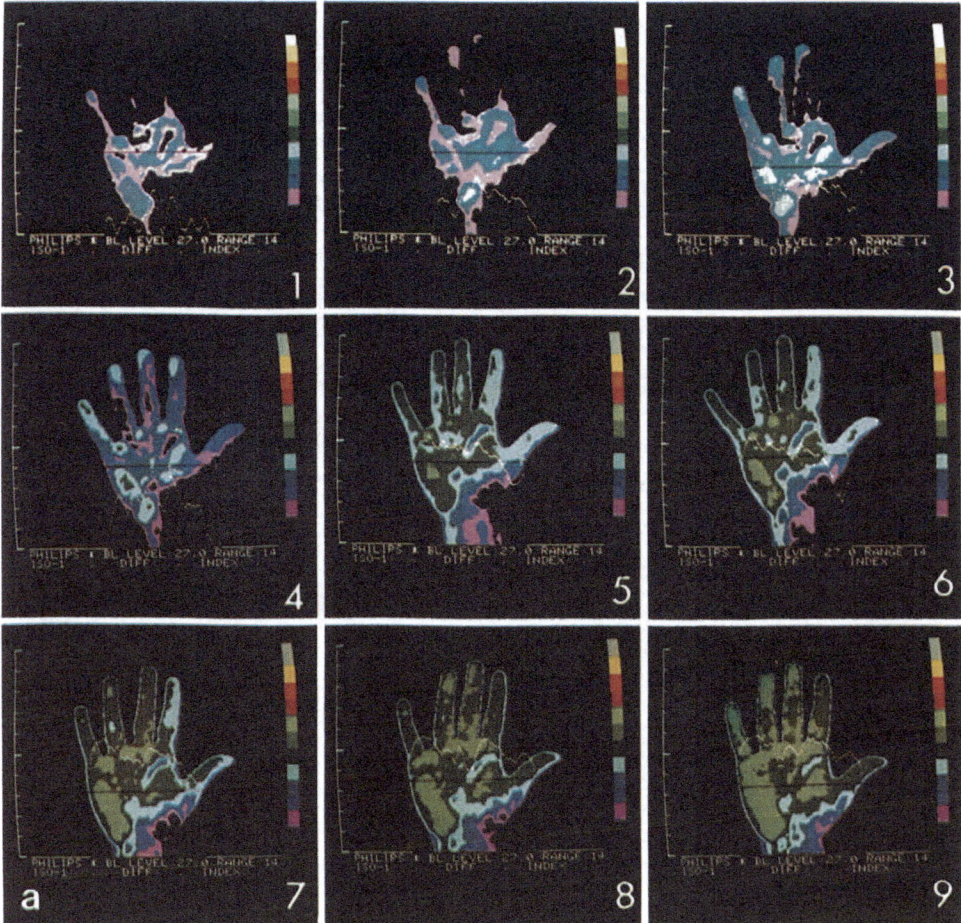

Fig. 12a.

Diagram (Fig. 11). Latency and regression of voluntary motor and vasomotor blockade were parallel and approximately simultaneous in ulnar and median regions. Latency time did differ by a few minutes between the voluntary (63 min) and vasomotor (70 min) systems. The ulnar part of the plexus had a markedly shorter diffusion distance – the sensations of cold and pain were blocked after 29 and 37 min, compared with 45 and 50 min for the median nerve. The latency time for the sensation of touch is in both cases identical with that for vasomotoricity (70 min). After washing out of the local anaesthetic, the median-innervated region is the first to show signs of regression. The sensation of touch returns unusually late.

Thermograms (Fig. 12a, b). Perfusion is started immediately after thermogram 1. Radiation of heat increases until thermogram 9 (70 min). The last changes in temperature can be recognized by the increase in size of the light-green area and the disappearance of the black level areas over the proximal part of the thenar. Thermograms 9–12 show the anaesthesia of the ulnar and median nerves in the "steady state" of total blockade. The partially blocked area

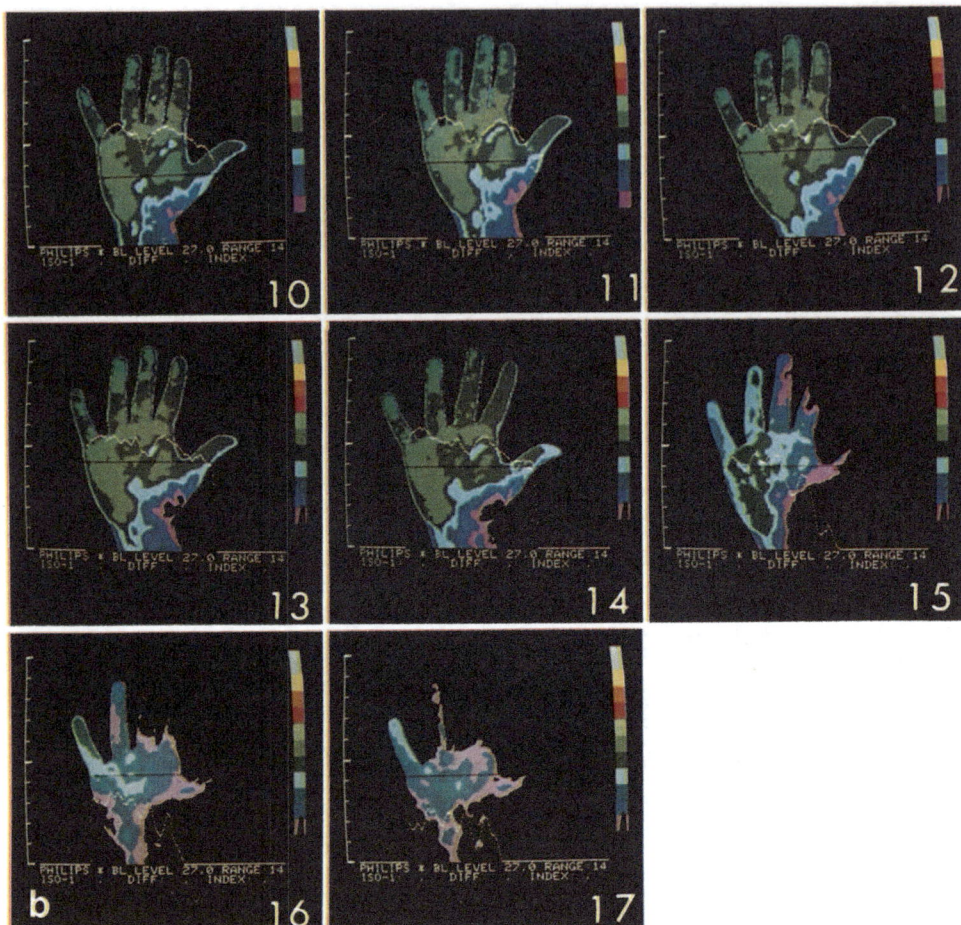

Fig. 12b.

over the proximal thenar is supplied partly by the median nerve and partly by the lateral cutaneous nerve of the forearm, but is nevertheless included in the blockade of the sensations of cold and pain. Thermograms 13–17 document the regression of the blockade, which plainly begins in the median region, where initial values are again reached.

Investigation 10

Brachial plexus blockade (latency and spontaneous regression)
Subject: 37-year-old man
Dose: 40 ml (1%) at 1 ml/min

Diagram (Fig. 13). Blockade developed irregularly in all the functions tested. In the region supplied by the *ulnar nerve,* voluntary motoricity was total after only 33 min, with the exception of minimal movement in the fifth finger. Between 45 and 60 min, motoricity re-

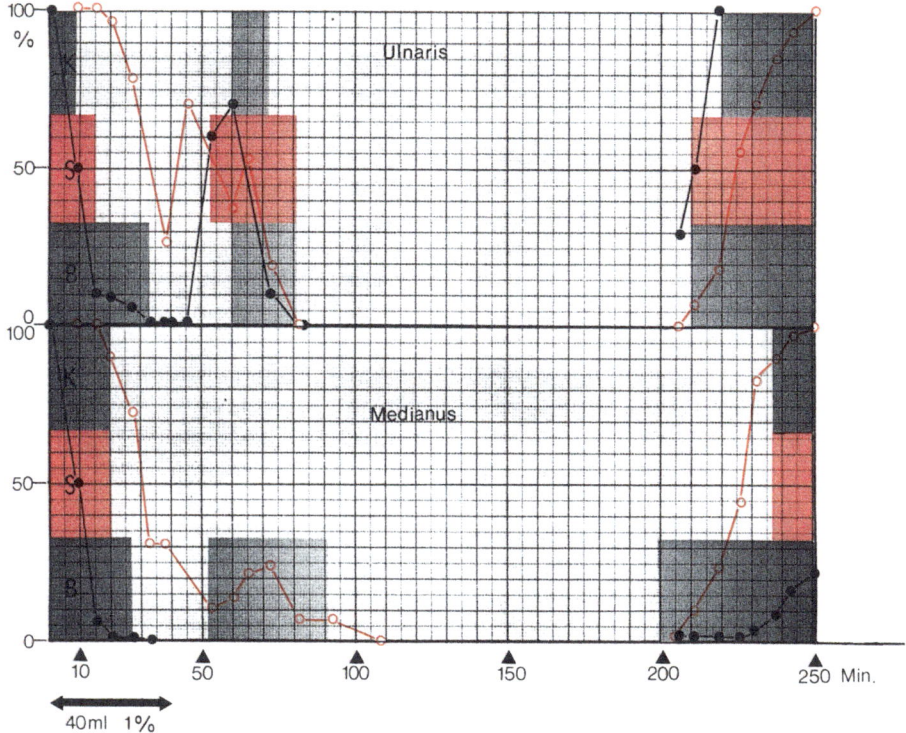

Fig. 13.

covered to 70% of initial values, but total blockade followed after 81 min. Vasomotor block-ade was later in beginning, but showed phases of regression during the same period and also became complete at 81 min. The sensations of cold, pain and touch also recovered tempora-rily during the regression phase. The final recovery of voluntary motoricity came 30 min be-fore that of vasomotoricity. In the area supplied by the *median nerve,* voluntary motoricity stayed blocked after the 33rd min, showing only slight regression during the period of the investigation. As in the ulnar region, vasomotoricity and the sensation of touch recovered briefly during the latency period. The final regression of vasomotor blockade was complete and simultaneous with that in the ulnar region.

Investigation 11

Brachial plexus blockade (latency and regression after washing out of the local anaesthe-tic)
Subject: 23-year-old woman
Dose: 40 ml (1%) at 1 ml/min

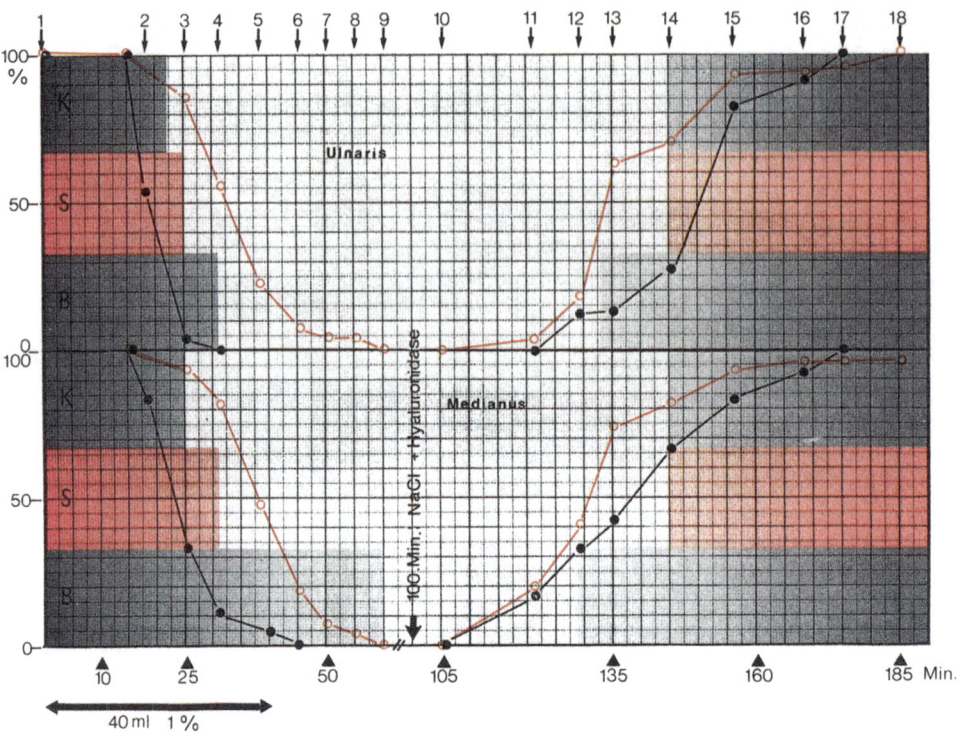

Fig. 14.

Diagram (Fig. 14). The voluntary motor and vasomotor blockades show different latency times. This is most marked in the ulnar supply region, where the voluntary blockade is total after 31 min, the vasomotor blockade only after 60 min have elapsed. Regression is essentially parallel and simultaneous. The sensation of touch is blocked together with voluntary motoricity in the ulnar region, together with vasomotoricity in the median region. As in all investigations, the sensations of cold and pain are blocked in quick succession during the phase in which the progression of voluntary motor blockade is most marked.

Thermograms (Figs. 15a, b and 16a, b) The simultaneously recorded colour and black-and-white show the index values over areas supplied by the median nerve and ulnar nerve respectively. Thermograms 1–9 (Fig. 15) display the measurements in the latency phase, thermograms 10–18 (Fig. 16) those in the regression phase. Again, perfusion commences immediately after thermogram 1. In thermogram 4 the black level is increased from 14.9 °C to 19.5 °C. Voluntary motor blockade in the ulnar supply area is already total at this point, but does not become total in the median supply area until thermogram 6. In thermogram 8 there is still a zone of clearly reduced radiation of heat over the third finger and third ray of the metacarpus, and the sensation of touch is preserved in exactly this zone. Thermograms 9 and 10 (60 and

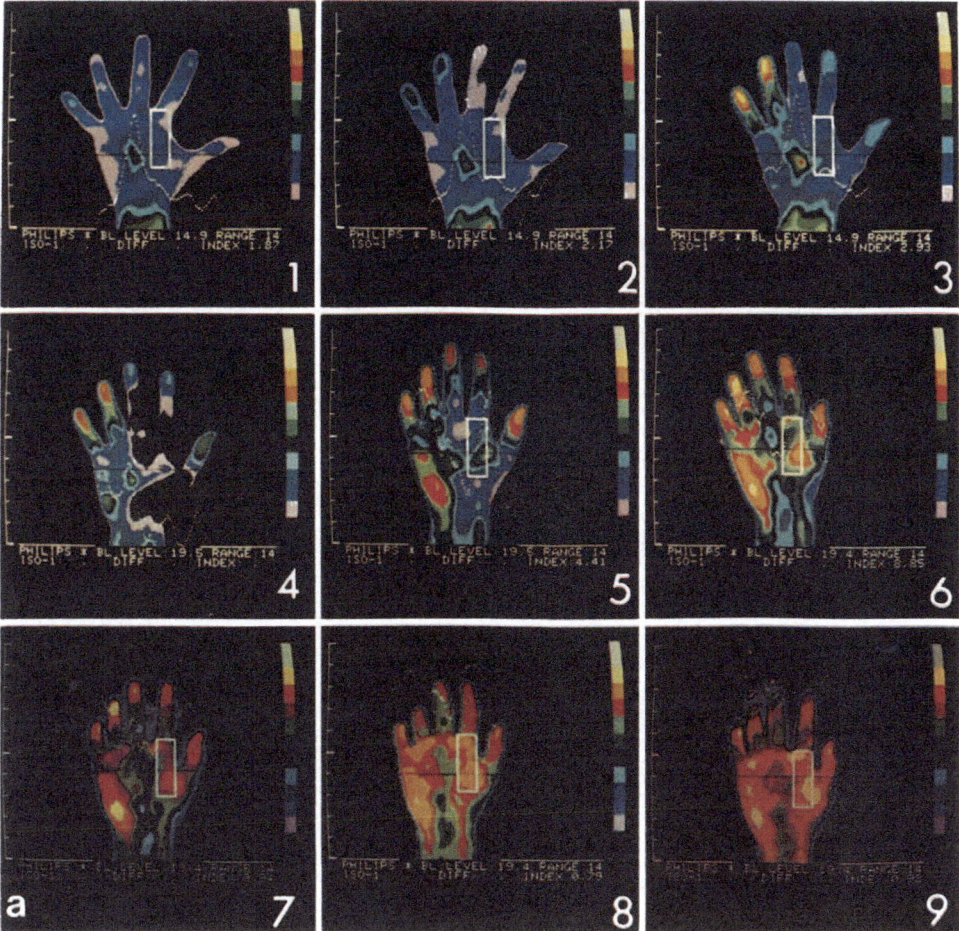

Fig. 15a.

105 min) show steady-state total blockade. Regression of the vasomotor blockade is essentially complete (finger temperature) by 165 min (thermogram 15). The mean temperatures continue to fall until thermogram 18 over the metacarpus, where the initial temperature is not completely reached.

Investigation 12

Tibial blockade (latency and spontaneous regression)
Subject: 25-year-old man
Dose: 20 ml (2%) at 1 ml/min

Diagram (Fig. 17). The course of voluntary motor and vasomotor blockade is simultaneous in both latency and regression phases, but the unfavourable position of the surface area

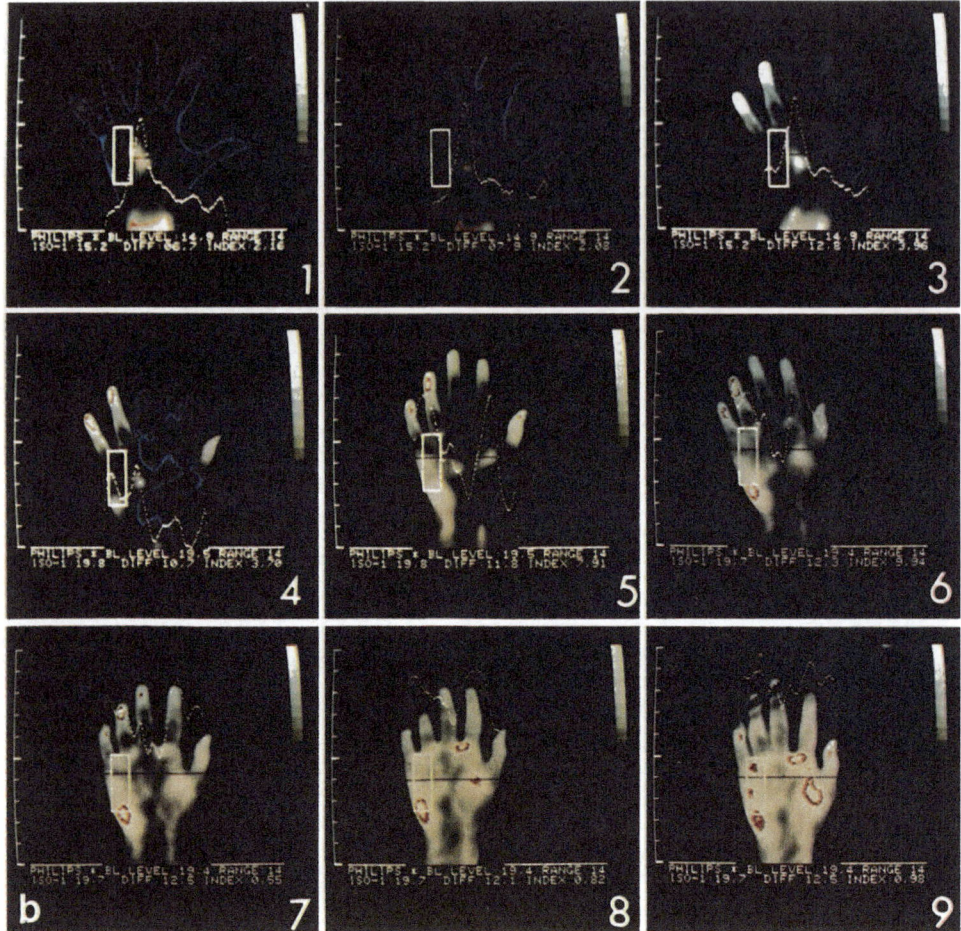

Fig. 15 b.

studied results in severe distortion of the curve of vasomotor blockade. The anaesthesia begins to affect voluntary motoricity 10 min before vasomotoricity, and both functions are totally blocked after 70 min, at which point the subject no longer feels touch stimuli. The sensations of cold and pain are blocked after 39 and 45 min respectively. Again, regression of blockade starts later in touch than in voluntary motoricity and vasomotoricity. Cold and pain stimuli are first felt, simultaneously, in the last 10 min of vasomotor blockade regression.

Investigation 13

Tibial blockade (latency and spontaneous regression)
Subject: 26-year-old man
Dose: 1) 20 ml (2%) at 1 ml/min; 2) 12 ml (2%) at 1 ml/min

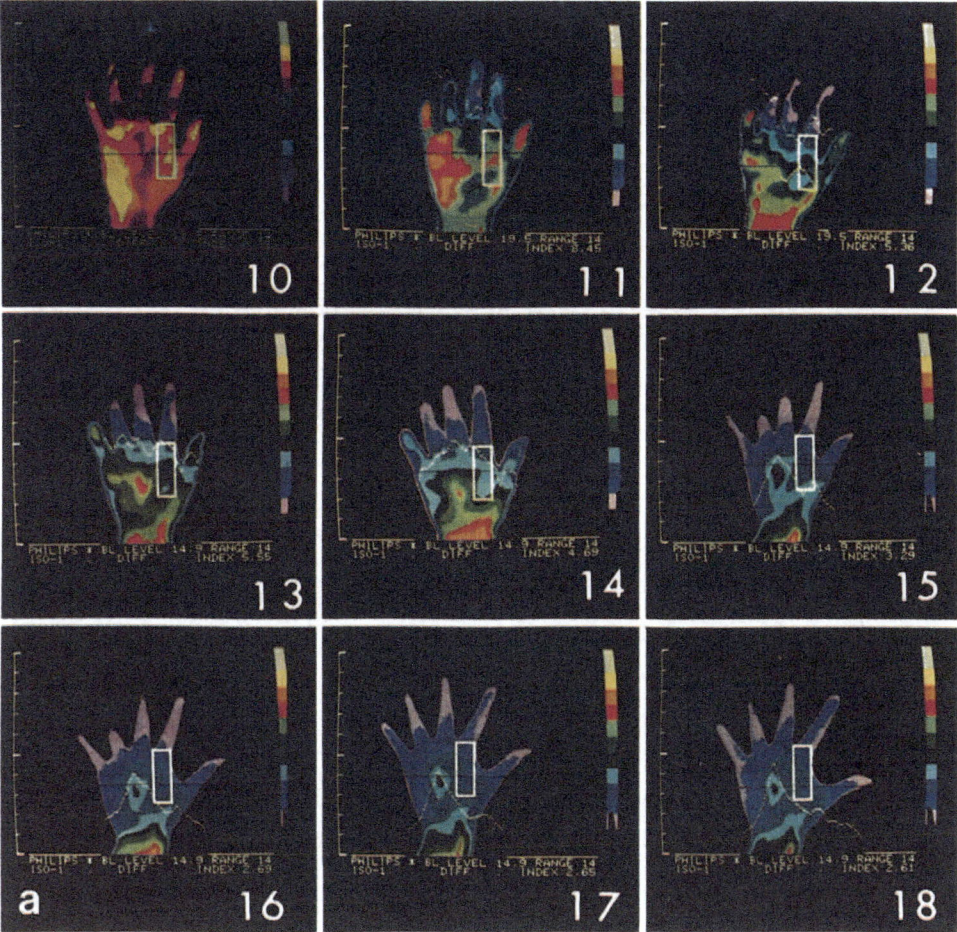

Fig. 16a.

Diagram (Fig. 18). The voluntary motor blockade begins long before the vasomotor blockade. Whereas vasomotoricity still shows no change after 30 min, voluntary motoricity is already 90% blocked. In the second phase of perfusion voluntary motoricity recovers by 50% and the incipient vasomotor blockade comes temporarily to a standstill. After 85 min, voluntary motoricity and the sensation of touch are totally blocked, and by this point vasomotor blockade is also essentially total, although minimal activity is still registered thermographically up to 100 min. There is no conduction of cold and pain stimuli after only 22 and 30 min respectively. Regression begins with the sensation of touch and runs practically simultaneously for voluntary motoricity and vasomotoricity. Pain stimuli are first perceived after 330 min, cold stimuli after 354 min.

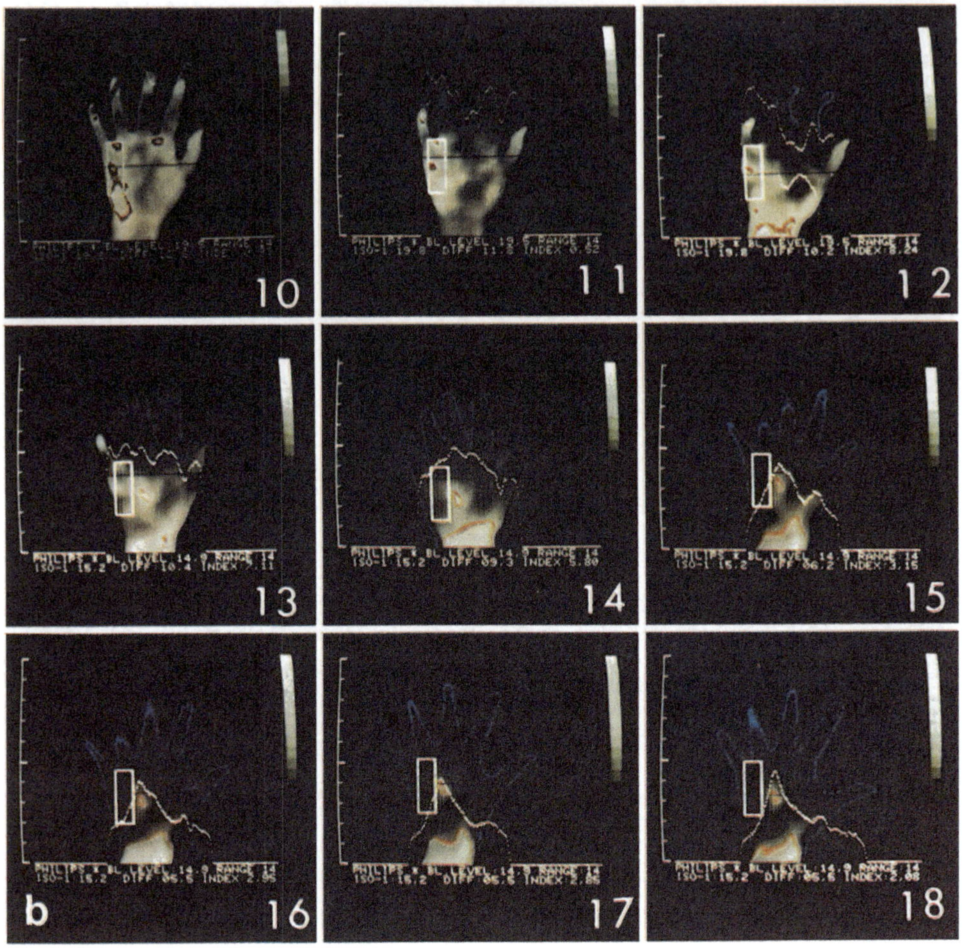

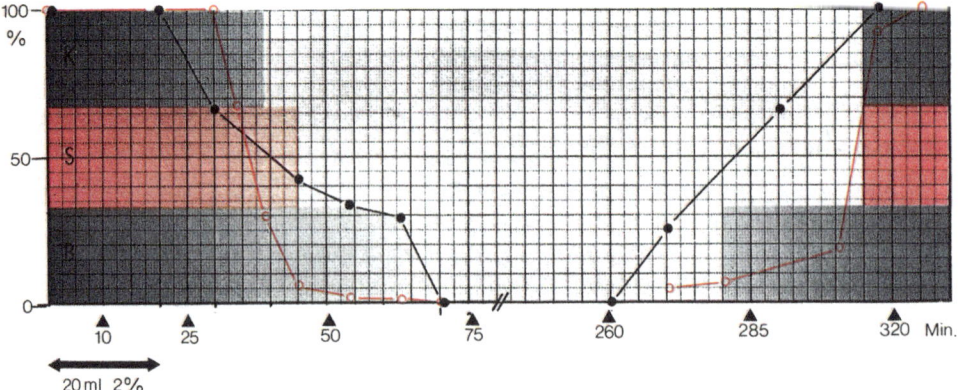

Fig. 16 b.

Fig. 17.

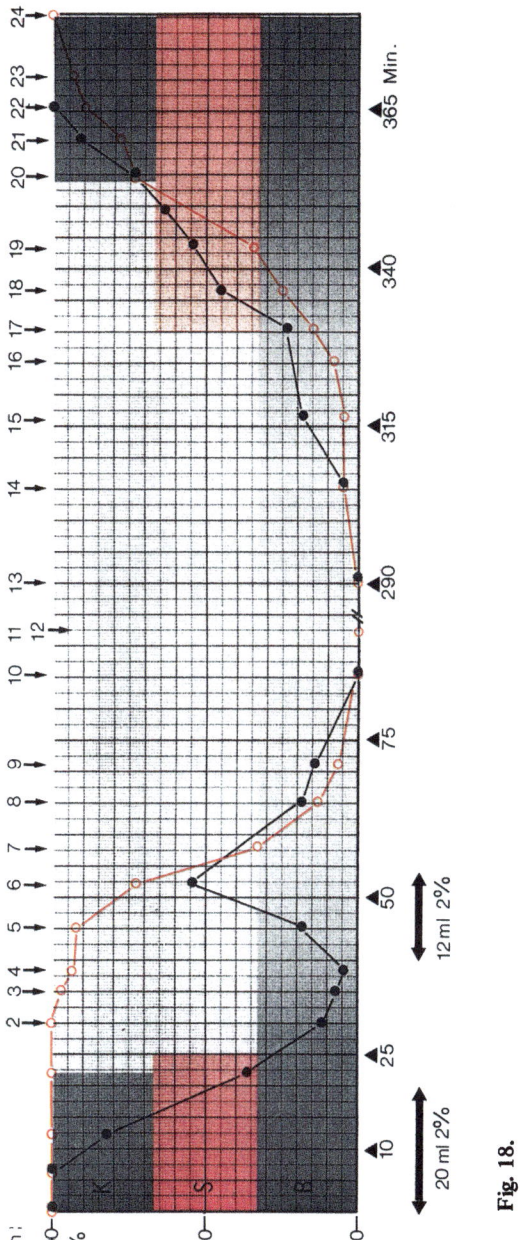

Fig. 18.

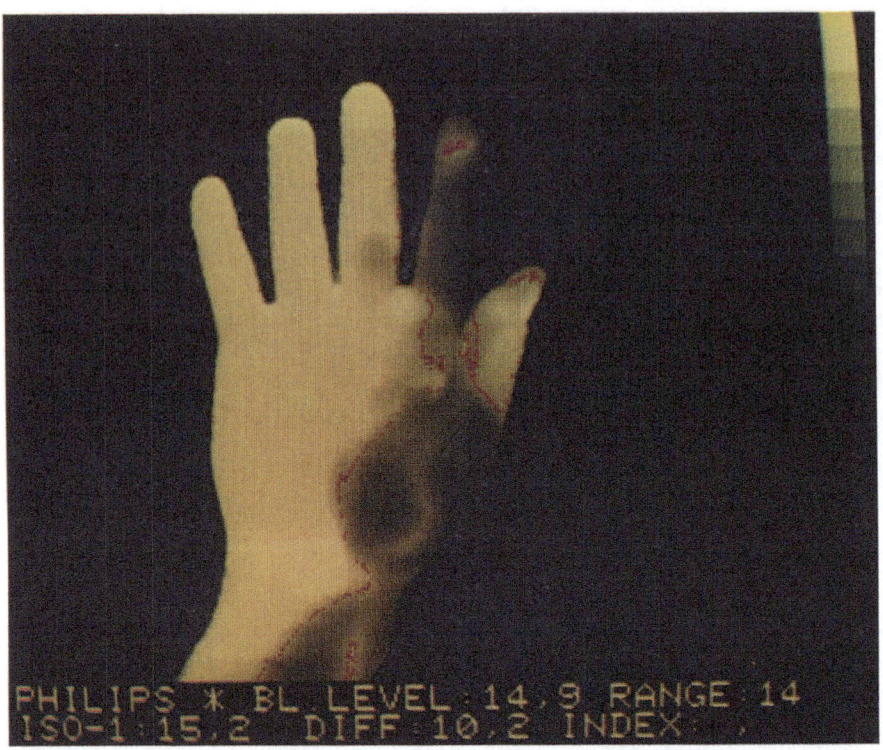

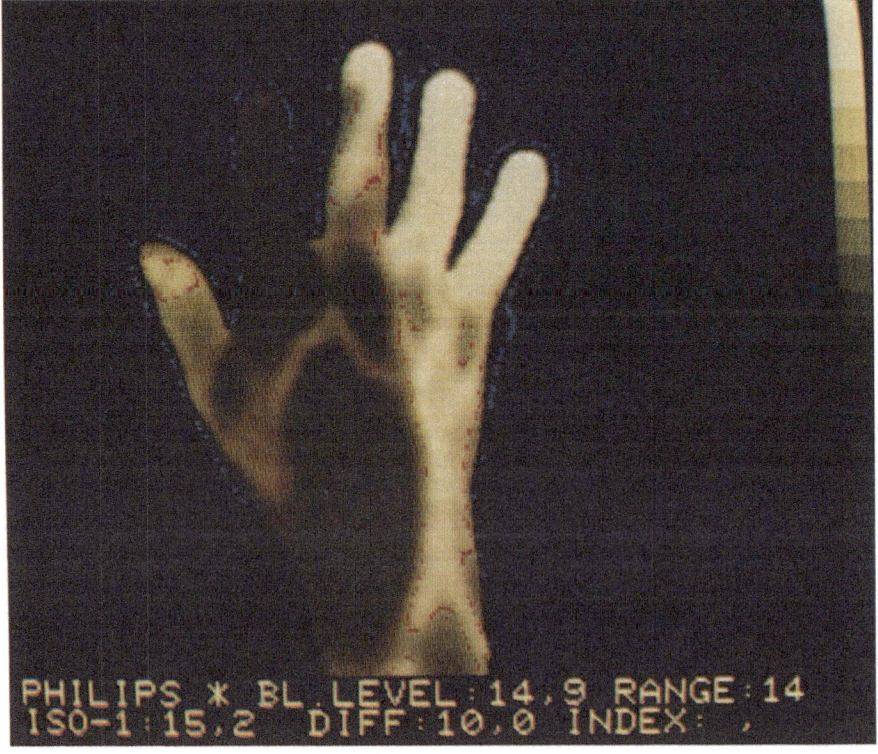

Fig. 19

Summary

The results of the individual investigations can be summarized as follows:

1. In none of the 13 investigations was vasomotoricity blocked before voluntary motoricity, and in Investigations 2, 3, 6, 10 and 11 it was blocked substantially later.
2. In most of the investigations, the first sign that the local anaesthetic was taking effect was a decrease in muscle power.
3. The blockade of the sensation of touch was usually simultaneous with that of voluntary motoricity and/or vasomotoricity.
4. The sensation of cold was blocked before or at the same time as that of pain. The loss of these sensations always occurred in the period of steepest decline in voluntary motor performance.
5. Despite the long duration of investigations, the onset of regression was in most cases simultaneous in vasomotoricity and voluntary motoricity. The regression of blockade was altogether more homogenous than the latency period.
6. Independently from the sensation of cold, the cooling in ice-water led to vasoconstriction until the blockade was total (Fig. 19).

◀ **Fig. 19.** Thermographic documentation of an intact cold reflex during complete blockade of the sensations of pain and cold. The two black-and-white thermograms show the volar (*upper*) and dorsal (*lower*) views of a hand in the "steady state" of an incomplete plexus anaesthesia. Voluntary motoricity in the reference muscles of the ulnar and radial nerves is completely blocked, and the flexion power of the index finger (median nerve) is only 10% of the initial value. Touch and pressure are only sensed over the clearly colder skin areas (*orange isotherm*), i.e. the innervation areas of the median nerve, the lateral cutaneous nerve of the forearm, and the superficial ramus of the radial nerve

Discussion

Criticism of Methods

Because of the intermittent cooling, continuous measurement over the investigation areas was not possible. There were therefore margins of variation in the length of the 4- to 5-min intervals between the measurements by which latency and regression times were defined. However, the protracted gradual administration of the local anaesthetic made the latency times so long that the range of error remained insignificant. The beginning and end of total voluntary motor blockade could always be recorded without measurements (cessation and recommencement of voluntary movements).

The findings in Investigations 11 and 13 reveal a further problem: With the investigations lasting several hours, it was not always possible to reconcile the minimal and maximal temperature during the latency and regression periods. In Investigation 13, regression was allowed to occur spontaneously, and measurements were discontinued for 190 min after total blockade was achieved. This long interruption of the intermittent cooling meant that when the investigation was recontinued, the temperature of the area being studied was higher at the onset or regression. In Investigation 11, the cooling was not interrupted, as rather than waiting for regression to occur spontaneously the local anaesthetic was washed out after 100 min. The temperature at the onset of regression was therefore not raised. While finger temperatures reached pre-investigation values as blockade faded (thermograms 16–18), the temperature in the region of the metacarpus remained above initial values. The point at which the regression of vasomotor blockade is complete therefore cannot be defined precisely.

One objection of principle to the chosen method of sympathetic stimulation must also be discussed: Muscle power in this experiment was only affected by the degree of nervo blockade because of uninfluenced central stimulation, whilst vasomotoricity was stimulated by a reflex circle the afferent limb of which was also affected by the nerve blockade. Activation of the two systems was therefore not completely comparable.

After the sensation of cold was blocked, one would have expected the cooling to be unable to produce vasomotor reflexes. According to the theory of differential block, it is at this point that conduction in fibres of the $A\delta$ group, and thus in the afferent limb of the vasomotor reflex, is interrupted. The high sensitivity of this group of fibres to local anaesthetics has been reported in all previous clinical-experimental studies [12, 22, 23, 42] and has been regarded as providing important support for this controversial theory. The loss of the sensation of cold, conducted by these fibres, also occurred long before motor blockade was complete in all the investigations in this study. While it is indisputable that stimulation of the cold reflex inevitable decreased as blockade increased, it was nevertheless firmly established that no clear decrease of this reflex activity could be observed on cessation of percep-

tion of cold. Up to the peak effect of the regional anaesthesia, the cooling in ice-water resulted in a clear separation of the variously intensively blocked skin areas in the thermogram
(Fig. 20, p. 38).

Discussion of Results

In none of the 13 investigations was the cold reflex, with its afferent conduction of impulses
via weakly myelinated Aδ fibres and efferent conduction via non-myelinated C fibres, blocked
before voluntary motoricity. This was independent of the concentration of local anaesthetic
and occurred with both short latency times (Investigation 7, 24 min) and long latency times
(Investigation 13, 85 min), as well as in steady-state partial blockade (Investigation 8). Only
the perception of cold, pain and touch stimuli was interrupted in "differential" fashion by
the regional anaesthesia. The loss of the sensation of cold without accompanying interruption of the cold reflex nevertheless shows that the blockade of perception in local anaesthesia
can by no means be equated with the blockade of conduction. The question of whether this
is also true for the perception and conduction of pain cannot be conclusively answered from
these experiments, but the analogy strongly suggests itself, since there is not even any mor-
phological support for differential sensitivity of pain fibres alone – skin pain stimuli are conducted by two different types of nerve fibres. Following the classification of Gasser and
Grundfest, these are Aδ fibres (Lloyd-Hunt group III) and non-myelinated C fibres (group
IV). These fibre types are morphologically identical to the afferent and efferent pathways
respectively of the cold reflex, for which no increased sensitivity to local anaesthetics compared with motor Aα fibres could be proved in these investigations. The results of Investigations 2, 3, 6, 10 and 11 even show clearly longer latency times for vasomotor blockade
than for voluntary motor blockade. This could be interpreted as confirming the findings of
Gissen et al. [26], who proved that postganglionic C fibres had the least sensitivity to local
anaesthetics and biotoxins. However, the lack of proof of a relatively increased latency time
in the other investigations and the almost simultaneous onset of regression in the fibres of
the voluntary motor and vasomotor systems in all investigations speak against such an inter-
pretation. Alone the different quantitative distribution of the fibre systems in these nerves
could simulate greater sensitivity of motor fibres if only latency times were taken into con-
sideration; on influx of the local anaesthetic, concentration gradients arise within the nerve,
thus there is a greater probability that the smaller proportion of motor fibres will be completely blocked. In regression of the blockade, on the other hand, lesser concentration gradients
are to be expected over the cross-section of the nerve. Before the onset of regression all fibres
are in the same condition of total blockade, and resorption takes place over shorter diffusion
distances through the network of fine capillaries of the blood vessels supplying the nerve.

The question posed at the outset of this study can therefore be answered unequivocally:
There is *no discernable connection* between the morphology of nerve fibres and their sensitivity to local anaesthetics. The latency times of the individual fibres in peripheral nerves are
determined solely by the concentration gradients of the dissociated local anaesthetic. The
dissociated loss of perception of cold, pain and touch cannot be explained by differential
sensitivity of the nerve fibres.

This finding contradicts all previous investigations of differential block. One reason for
this lies in the different methods employed to quantify the sympathetic blockade. Another

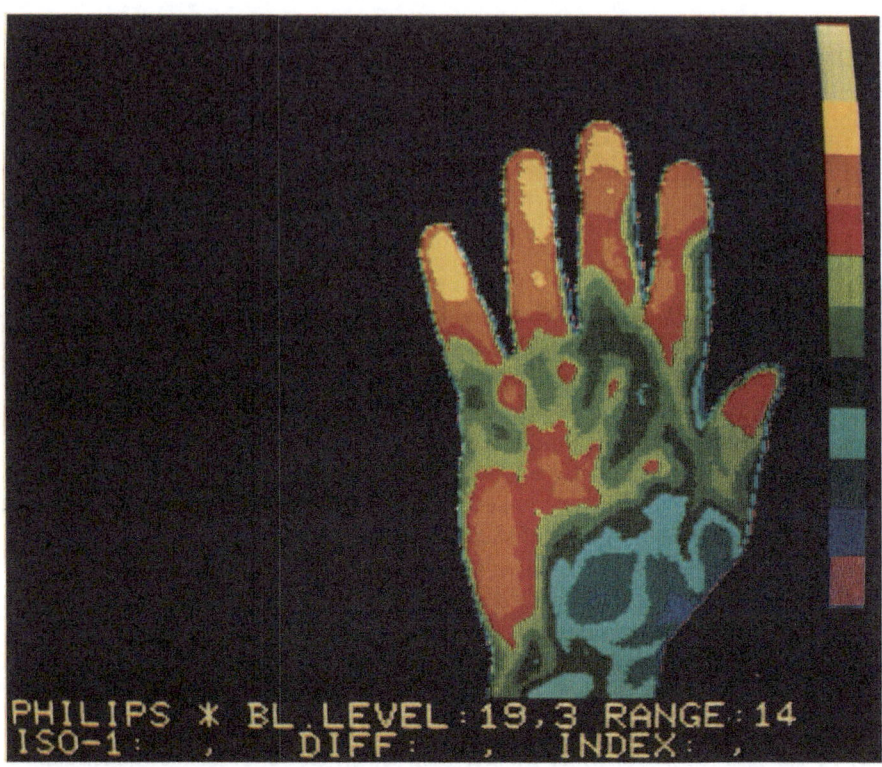

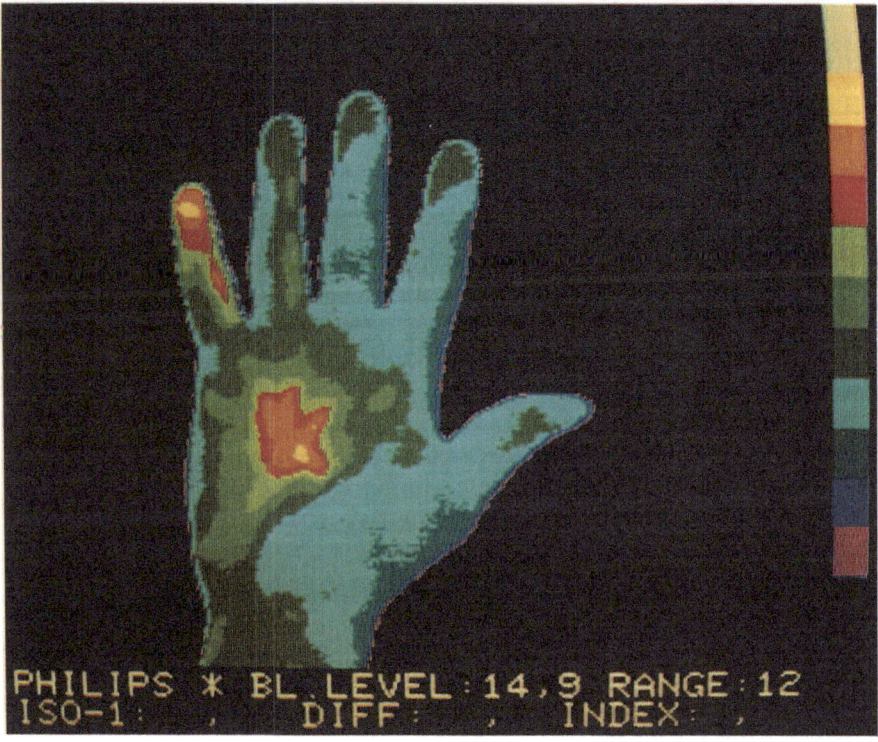

Fig. 20

reason is that the research teams centred on Nolte [42–44, 62] and Fruhstorfer [22], the only ones to carry out comparable clinical investigations, assumed, in interpreting their results, the validity of the previously undisputed hypothesis that the loss of perception of a sensation in regional anaesthesia can be equated with the interruption of conduction. Fruhstorfer et al. [22] went even further with their confidence in the hypothesis; when they found a dissociated blockade of the sensations of heat and cold in the ulnar nerve, they concluded that these sensations are conducted by morphologically different nerve fibres.

Harmes [29] and Greene [27] also presupposed the validity of this hypothesis in their investigations into spinal and extradural anaesthesia. The theory of differential block is therefore no longer founded on clinically relevant findings.

A Notable Incidental Finding: Influence of Further Injections on Dynamics of Blockade

During the latency period of Investigations 1, 2, 4, 6, 8 and 13, the perfusion with local anaesthetic was either accelerated (Investigation 6) or recontinued after an interval of variable length in order to bring about a total blockade. This led temporarily to phenomena – clearly dependent on the perfusion rate chosen, the volume of solution perfused and the previously reached degree of saturation of the fibre groups – which did not conform to the expected acceleration of anaesthesia. Small volumes of solution (Investigations 1 and 2) resulted only in a brief standstill, while greater volumes (Investigation 13) or a doubled perfusion rate (Investigation 6) caused a marked regression of the blockade. In Investigation 13, voluntary motoricity, for example, recovered by 50%, while the vasomotor blockade, which had only just begun, only came to a brief standstill. In contrast, in Investigation 6, vasomotoricity recovered to the initial value, whereas voluntary motor blockade regressed by 10% after a slight delay. Such reactions to later second injections have never before been described, and their interpretation is made even more difficult by the fact that the same phenomena occurred in Investigation 10 without a second injection.

The dynamics of the nerve blockade are determined during the latency period by the concentration gradients of the free base between extracellular space and nerve membrane. A further injection or an acceleration of perfusion must therefore bring about a reversal of the concentration gradients in order to cause the events described. This can readily be imagined, as the increased amount of inflowing local anaesthetic still contains no free base, and can, through its acid environment, convert part of the dissociated local anaesthetic in the tissue back into the salt form and considerably dilute the rest. In this way a reversal of the concen-

---

◀ Fig. 20. The significance of cooling in the quantitative assessment of vasomotor blockade by thermography. Brachial plexus blockade in the regression phase: complete analgesia, no sensation of cold, motoricity and sensation of touch present. (*Upper*): Without cooling, the hand shows relatively homogenous radiation of heat over the volar surface of all the fingers and the metacarpus. Only the terminal distribution area of the lateral cutaneous nerve of the forearm is clearly colder. (*Lower*): The same hand 2 min later after cooling in ice water: there is considerable cooling through vasoconstriction, and the clear temperature differences between the ulnar and the radial (median-innervated) palm show differing depths of blockade in these regions

tration gradients can be brought about under conditions of high membrane concentrations during partial blockade. The sometimes divergent effects on voluntary motoricity and vaso-motoricity show that the blockades of these fibres were at different stages. This indirectly confirms that the measurements made given a realistic picture of the dynamics.

The course of Investigation 10 constitutes only an apparent exception under this inter-pretation of the observations. It is conceivable that free movement of the arm led to tempo-rary blockage of the flow through the cannula or the neurovascular sheath and to enclosure of the local anaesthetic under high pressure in a small space. Accidental opening of this "valve" through a movement of the shoulder must then have allowed a renewed influx of local anaesthetic to the parts of the plexus previously cut off from perfusion. The strikingly short latency time of voluntary motor blockade in the median nerve supply region indicates pressure damage to the nerve, which must have occurred during the blockage of flow. The strongly pressure-sensitive motor fibres of this nerve recovered only slightly during the whole period of regression of vasomotor blockade. For several days subsequently the subject com-plained of a circumscribed partial paresis of the nerve.

Summary

The fibres of the peripheral and autonomic nervous system are divided into well-defined groups according to their conduction speed and physiological function. The groups are differentiated by axonal diameter and degree of myelination. According to a theory first formulated over 50 years ago, they are also distinguished from one another by differential sensitivity to local anaesthetics. The theory states that sensitivity decreases with increasing axonal diameter and myelination. The use of a sufficiently low concentration of local anaesthetic leads to the so-called differential block, not only in the recording of summation potentials in vitro, but also in clinical regional anaesthesia of peripheral nerves. The differential block consists in isolated interruption of conduction in non-myelinated fibres of the autonomic nervous system and in only slightly myelinated afferent fibres of temperature and pain receptors while voluntary motoricity and the sensation of touch remain unaffected. In regional anaesthesia using clinical doses, the differential sensitivity leads to a marked dissociation of latency times in these fibre systems. The theory of differential block has become a fundamental principle of neurophysiology, pharmacology and anaesthesiology and has found expression in all the textbooks in these fields. Clinically, the differential block is used for differentiation, localization and sometimes therapy of vegetative pain syndromes, and also in obstetric peridural anaesthesia, in order to achieve analgesia of labour pains without affecting voluntary motoricity [6, 56, 58, 63].

Doubts have been expressed regarding the experimental foundations of the theory ever since its formulation, but in the face of the convincing agreement of the neurophysiological experiments with the clinical data, have never led to a revision of it.

The thermographic documentation of therapeutic differential blockades and clinical-dose regional anaesthesias led eventually to results which cast doubt on at least the clinical relevance, if not even the validity of the theory [57]. Therefore, regional anaesthesia of peripheral nerves was conducted under standardized conditions in 13 subjects in order to quantify the course of blockade of the various nerve fibre groups. The local anaesthetic employed was mepivacaine, an agent with a medium duration of effect, and it was administered slowly and mechanically in various concentrations. The latency and regression times of blockade for voluntary motor and vasomotor (sympathetic) activity and for the sensations of cold, pain and touch were recorded in the median nerve, the tibial nerve and the brachial plexus. The standard for voluntary motoricity was the maximal power of a reference muscle supplied by the nerve in question; the power was measured in terms of the pull exerted on a spring balance by the bending of fingers or the great toe. The activity of the sympathetically innervated vasomotor system was determined by telethermographic measurement of the heat radiated over the autonomous area of the nerve concerned in the region of the palmar surface of the hand or the plantar surface of the foot. In order to measure vasomotor performance in such a way as to allow comparison with voluntary motoricity, the vasomotor reflex was maximally stimulated by ice-water cooling of the hands and feet under investigation. The investigations yielded the following results:

The solely sympathetically innervated vasomotor system was in every case totally blocked at the same time as or after voluntary motoricity and almost always simultaneously with the sensation of touch. The regression of vasomotor and voluntary motor blockade also began simultaneously in almost all investigations. The sensations of cold and pain were regularly blocked before the sensation of touch, voluntary motoricity and vasomotoricity, and regression of blockade also commenced later. Even after early blockade of the sensation of cold, the cold reflex remained unaffected. The ice-water cooling produced measurable vasoconstriction right up until blockade of the nerve became total, and the reflex returned as soon as regression began. The dissociated loss of sensations in regional anaesthesia thus does not rest on differential sensitivity of nerve fibres. It seems obvious that this dissociation takes place on a more highly organized level of the central nervous system.

These investigations in human subjects have furnished indisputable proof that there is no clinically relevant association between the morphology of nerve fibres and their sensitivity to local anaesthetics.

References

1. Altura BM, Altura BT (1974) Effects of local anesthetics, antihistamines, and glucocorticoids on peripheral blood flow and vascular smooth muscle. Anesthesiology 41:197
2. Barnes RB, Gerson-Cohen J (1963) Clinical thermography. JAMA 185:949
3. Barnes RB (1963) Thermography of the human body. Science 140:870
4. Bier A (1899) Versuche über die Cocainisierung des Rückenmarkes. Dtsch Z Chir 51:361
5. Blaustein MP, Goldman DE (1966) Competitive action of calcium and procaine on lobster axon. J Gen Physiol 49:1043
6. Bonica JJ (1953) Management of pain. Lea & Febiger, Philadelphia
7. Brade R (1966) Vergleichende Untersuchungen über die Wirksamkeit von Mepivacain und dem neuen Lokalanaesthetikum Marcain (Lac 43). Diss. Univ. Mainz
8. Brill S, Lawrence LB (1930) Changes in temperature of the lower extremities following the induction of spinal anesthesia. Proc Soc Exper Biol Med 27:728
9. Büchi J, Perlia X (1960) Beziehungen zwischen den physikalisch-chemischen Eigenschaften und der Wirkung von Lokalanaesthetika. Arzneimittelforsch 10:177
10. Chen RYZ, Lee MML, Chien S (1979) Local anesthetics and the rheologic behaviour of erythrocyte suspensions. Anesthesiology 51:245
11. Covino BG, Vasallo HG (1976) Local anesthetics, mechanism of action and clinical use. Grune & Stratton, New York
12. De Jong RH, Wagman IH (1963) Physiological mechanisms of peripheral nerve block by local anesthetics. Anesthesiology 24:684
13. De Jong RH (1980) Differential nerve block by local anesthetics. Anesthesiology 53:443
14. Dhuner KG, Lewis DH (1966) Effect of local anesthetics and vasoconstrictors upon regional blood flow. Acta Anaesth Scand Suppl 23:352
15. Dittmann ECH, Zipf HF (1973) Schmerzauslösung und Pharmakologie der Lokalanaesthetika. In: Killian H (Hrsg) Lokalanaesthesie und Lokalanaesthetika. Thieme, Stuttgart
16. Dixon WE (1905) Selective action of cocaine on nerve fibers. J Physiol 32:87
17. Douglas WW, Ritchie JM (1962) Mammalian non myelinated nerve fibers. Physiol Rev 42:297
18. Everett GM, Goodsell JS (1954) Greater resistance to procaine of slow fiber groups in some peripheral nerves. J Pharmacol 106:385
19. Everett GM, Toman JEP (1954) Procaine block of fiber groups in various nerves. Fed Proc 13:385
20. Folkow B (1952) Impulse frequency in sympathetic vasomotor fibers correlated to the release and elimination of the transmitter. Acta Physiol Scand 25:49
21. Franz DM, Perry RS (1974) Mechanisms for differential block among single myelinated and non myelinated axons by procaine. J Physiol 236:193
22. Fruhstorfer H, Zenz M, Nolte H, Hensel H (1974) Dissociated loss of cold and warm sensibility during regional anaesthesia. Pflügers Arch 349:73
23. Fruhstorfer H (1977) Differenzierte Ulnarisblockade mit mittellang- und langwirkenden Lokalanaesthetika. In: Meyer J, Nolte H (eds) Die Pharmakologie, Toxikologie und klinische Anwendung langwirkender Lokalanaesthetika. Thieme, Stuttgart, p 24
24. Gasser HS, Erlanger J (1929) The role of fiber size in the establishment of a nerve block by pressure or cocaine. Amer J Physiol 88:581
25. Gasser HS, Grundfest H (1939) Axon diameters in relation to the spike dimensions and the conduction velocity in mammalian A-fibers. Amer J Physiol 127:393

26. Gissen AJ, Covino BG, Gregus J (1980) Differential sensitivities of mammalian nerve fibers to local anaesthetic agents. Anesthesiology 53:467
27. Greene NM (1955) Area of differential block in spinal anesthesia with hyperbaric tetracaine. Anesthesiology 16:573
28. Greenfield ADM, Shepherd JZ (1950) A quantitative study of the response to cold of the circulation through the fingers of normal subjects. Clin Sc 9:323
29. Harmes R (1978) Differentielle Blockade (sympathisch-sensibel) bei Peridural- und Spinalanaesthesie. Diss. Univ. Mainz
30. Heavner JE, de Jong RH (1974) Lidocaine blocking concentration for B and C nerve fibers. Anesthesiology 40:228
31. Heinbecker P, Bishop GH, O'Leary J (1934) Analysis of sensation in terms of nerve impulse. Arch Neurol Psychiat 31:34
32. Jorfeldt L, Löfström B, Pernow B, Wahren J (1970) The effect of mepivacaine and lidocaine of forearm resistance and capacitance vessels in man. Acta Anaesthesiol Scand 14:183
33. Kendig JJ, Cohen EN (1977) Pressure antagonism to nerve conduction block by anesthetic agents. Anesthesiology 47:6
34. Lee AG (1976) Model for action of local anaesthetics. Nature 262:545
35. Lewis LW (1963) Sympathetic nerve block. Int Anesth Clin 1:613
36. Lloyd DPS, Chang HT (1948) Afferent fibers in muscle nerves. J Neurophysiol 11:199
37. Matthews PBC, Rushworth G (1957) The relative sensitivity of muscle nerve fibers to procaine. J Physiol 135:263
38. Narahashi T, Frazier DT (1971) Site of action and active form of local anesthetics. Neurosci Res 4:65
39. Nathan PW, Sears TA (1961) Some factors concerned in differential nerve block by local anaesthetics. J Physiol 157:565
40. Nathan PW, Sears TA (1962) Differential nerve block by sodium-free and sodium-deficient solutions. J Physiol 164:375
41. Nathan PW, Sears TA (1963) The suceptibility of nerve fibers to analgesics. Anaesthesia 18:467
42. Nolte H, Meyer J, Köpf B, Zenz M (1974) Klinische und elektrophysiologische Parameter zur Differenzierung der Wirkung von Lokalanaesthetika. Anaesthesist 23:165
43. Nolte H, Fruhstorfer H, Pfaff U, Radtke J, Zenz M (1976) Dissociation of cold, warm and hot sensibility during ulnar nerve block and surface anesthesia. In: Bonica JJ (ed) Advances in pain research and therapy. Raven Press, New York, p 673
44. Nolte H, Farrar MD (1978) Upper arm blocks: differential modalities. In: Stanton-Hicks, Md'A (ed) Regional Anesthesia: Advances and selected topics. IAC 16/4:183
45. Nuhn P, Frenzel J, Arnold K (1979) Zur Wechselwirkung von Lokalanaesthetika und Neuroleptika mit Membranen. Pharmazie 34/3:131
46. Ritchie JM, Greengard P (1966) On the mode of action of local anesthetics. Ann Rev Pharmacol 6:405
47. Rogwalder H (1968) Vergleichende Untersuchungen über die Latenz-, Wirkungs- und Regressionszeit der Lokalanaesthetika Mepivacain und Bupivacain. Diss Univ. Mainz
48. Rosenberg PH, Heinonen E, Jansson SE, Gripenberg J (1980) Differential nerve block by bupivacaine and 2-chloroprocaine. An experimental study. Br J Anaesth 52:1183
49. Schoop W (1959) Physiologie und Pathophysiologie der peripheren Durchblutung. In: Ratschow M (ed) Angiologie. Thieme, Stuttgart, p 86
50. Seeman P (1972) The membrane actions of anesthetics and tranquilizers. Pharmacol Rev 24:583
51. Shanes AM (1958) Electrophysiological aspects of physiological and pharmacological action in excitable cells. Pharmacol Rev 10:59
52. Shepherd JT (1964) Regulation of blood flow in human limbs. In: Price HL, Cohen PJ (ed) Effects of anesthetics on the circulation. Charles C Thomas, Springfield, p 262
53. Sinclair DC, Hinshaw JR (1950) Sensory changes in procaine nerve block. Brain 73:224
54. Skou JC (1958) Relation between the ability of various compounds to block nerve conduction and their penetration into a monomolecular layer of nerve tissue lipoids. Biochem Biophys Acta 30:625
55. Spealman CR (1945) Effect of ambient air temperature and of hand temperature on blood flow in hands. Amer J Physiol 145:218
56. Sprotte G (1977) Erfahrungen mit der differenzierenden epiduralen Blockade in der Geburtshilfe. Anaesthesiologische Informationen 1977/4:180

57. Sprotte G (1980) Telethermographische Beobachtungen bei der Ausbreitung rückenmarksnaher Leitungsanaesthesien. Anaesth Intensivmed 130:319
58. Sprotte G, Rietbrock I, Lehmann V, Roebke A (1981) Differenzierende peridurale Analgesie für die vaginale Entbindung. Regionalanaesthesie 4:49
59. Strichartz G (1976) Molecular mechanisms of nerve block. Anesthesiology 45:421
60. Tasaki I (1953) Nervous transmission. Charles C Thomas, Springfield (Ill)
61. Toman JEP (1952) Neuropharmacology of peripheral nerve. Pharmacol Rev 4:168
62. Wencker KH, Nolte H, Fruhstorfer H (1975) Brachial plexus blockade for evaluation of local anaesthetics. Br J Anaesth 47:301
63. Winnie AP, Collins VJ (1968) The pain clinic I. Differential neural blockade in pain syndromes of questionable etiology. Med Clin North Am 52:123
64. Zipf HF (1967) Die Wirkungsmechanismen der Lokalanaesthetika. Pharm Acta Helv 42:480

Anaesthesiologie und Intensivmedizin

Anaesthesiology and
Intensive Care Medicine

vormals „Anaesthesiologie und Wiederbelebung"
begründet von R. Frey, F. Kern und O. Mayrhofer

Herausgeber: H. Bergmann (Schriftleiter)
J. B. Brückner, M. Gemperle, W. F. Henschel,
O. Mayrhofer, K. Meßmer, K. Peter

Band 145
J. Beyer, K. Messmer
Organdurchblutung und Sauerstoffversorgung bei PEEP
Tierexperimentelle Untersuchungen zur regionalen
Organdurchblutung und lokalen Sauerstoffversorgung
bei Beatmung mit positiv-endexspiratorischem Druck
1982. 17 Abbildungen, 18 Tabellen. X, 84 Seiten
Broschiert DM 54,-. ISBN 3-540-11220-0

Band 147
L. Tonczar
Kardiopulmonale Wiederbelebung
1982. 44 Abbildungen, 15 Tabellen. XIII, 143 Seiten
Broschiert DM 64,-. ISBN 3-540-11760-1

Band 148
Regionalanaesthesie
Ergebnisse des Zentraleuropäischen Anaesthesie-
kongresses Berlin 1981
Band 1
Herausgeber: J. B. Brückner
1982. 125 Abbildungen, 43 Tabellen. XIII, 215 Seiten
Broschiert DM 83,-. ISBN 3-540-11744-X

Band 149
Inhalationsanaesthesie heute und morgen
Herausgeber: K. Peter, F. Jesch
Übersetzungen aus dem Englischen
von E. Mertens-Feldbausch
1982. 126 Abbildungen, 19 Tabellen. XII, 276 Seiten
Broschiert DM 42,-. ISBN 3-540-11756-3

Band 150
Inhalation Anaesthesia Today and Tomorrow
Editors: K. Peter, F. Jesch
1982. 126 figures. 272 pages
Soft cover DM 76,-. ISBN 3-540-11757-1

Band 151
H. Marquort
Kontraktionsdynamik des Herzens unter Anaesthetika und Beta-Blockade
Tierexperimentelle Untersuchungen
1983. 137 Abbildungen, 34 Tabellen. XVI, 202 Seiten
Broschiert DM 62,-. ISBN 3-540-11745-8

Band 152
Der Anaesthesist in der Geburtshilfe
Ergebnisse des Zentraleuropäischen Anaesthesie-
kongresses, Berlin 1981
Band 2
Herausgeber: J. B. Brückner
1982. 68 Abbildungen, 19 Tabellen. X, 184 Seiten
Broschiert DM 46,-. ISBN 3-540-11831-4

Band 153
Schmerzbehandlung – Epidurale Opiatanalgesie
Ergebnisse des Zentraleuropäischen Anaesthesie-
kongresses Berlin 1981
Band 3
Herausgeber: J. B. Brückner
1982. 90 Abbildungen, 50 Tabellen.
XII, 194 Seiten (24 Seiten in Englisch)
Broschiert DM 74,-. ISBN 3-540-11830-6

Band 154
R. Larsen
Kontrollierte Hypotension
Durchblutung und Sauerstoffverbrauch des Gehirns
und des Herzens
1983. 20 Abbildungen, 19 Tabellen. VII, 88 Seiten
Broschiert DM 35,-. ISBN 3-540-11921-3

Band 155
K. Inoue
Vagaler Herztonus und Herzfrequenz unter dem Einfluß von Injektionsanaesthetika
Eine Studie an narkotisierten Katzen
1983. 11 Abbildungen, 3 Tabellen. IX, 39 Seiten
Broschiert DM 24,-. ISBN 3-540-12031-9

Springer-Verlag
Berlin
Heidelberg
New York
Tokyo

Anaesthesiologie und Intensivmedizin

Anaesthesiology and Intensive Care Medicine

vormals „Anaesthesiologie und Wiederbelebung"
begründet von R. Frey, F. Kern und O. Mayrhofer

Herausgeber: H. Bergmann (Schriftleiter),
J. B. Brückner, M. Gemperle, W. F. Henschel,
O. Mayrhofer, K. Meßmer, K. Peter

Band 156
Hämodynamisches Monitoring
Workshop Erbach 14. Mai 1982
Herausgeber: F. Jesch, K. Peter
1983. 97 Abbildungen, 20 Tabellen. VI, 170 Seiten
Broschiert DM 62,-. ISBN 3-540-12093-9

Band 157
Kinderanaesthesie
Prämedikation – Narkoseausleitung
Ergebnisse des Zentraleuropäischen Anaesthesie-
kongresses Berlin 1981
Band 4
Herausgeber: J. B. Brückner
1983. 162 Abbildungen, 75 Tabellen. XIII, 275 Seiten
Broschiert DM 108,-. ISBN 3-540-12153-6

Band 158
Neue Aspekte in der Regionalanaesthesie 3
Plexus- und Epiduralanaesthesie: Technik und
Komplikationen
Opiate epidural, intrathekal
Herausgeber: H. J. Wüst, M. D'Arcy Stanton-Hicks,
M. Zindler
1984. 113 Abbildungen, 67 Tabellen. XV, 250 Seiten.
Broschiert DM 98,-. ISBN 3-540-13023-3

Band 160
H. Goslinga
Blood Viscosity and Shock
The Role of Hemodilution, Hemoconcentration and
Defibrination
1984. 79 figures, 4 tables. XXVI, 193 pages
Soft cover DM 78,-. ISBN 3-540-12620-1

Band 161
Deutscher Anaesthesiekongreß (DAC 82)
Freie Vorträge
2.-6. Oktober 1982 in Wiesbaden
Herausgeber: J. Schara
1984. 236 Abbildungen, 107 Tabellen.
XVIII, 393 Seiten
Broschiert DM 158,-. ISBN 3-540-12977-4

Band 162
G. Meuret
Pharmakotherapie in der Reanimation nach Herz-Kreislauf-Stillstand
Untersuchungen an Hunden und an isolierten
Meerschweinchenherzen
1984. 55 Abbildungen, 10 Tabellen. XVII, 116 Seiten
Broschiert DM 78,-. ISBN 3-540-12978-2

Band 163
D. Scheidegger, L. J. Drop
Ionisiertes Kalzium
Seine Messungen und seine kardiovaskulären
Auswirkungen
1984. 30 Abbildungen, 3 Tabellen. X, 57 Seiten
Broschiert DM 34,-. ISBN 3-540-13567-7

Band 164
Das Berufsbild des Anaesthesisten
Herausgeber: J. B. Brückner, P. Uter
1984. 21 Abbildungen, 26 Tabellen. X, 168 Seiten
Broschiert DM 68,-. ISBN 3-540-13467-0

Band 167
Intensive Care and Emergency Medicine
4th International Symposium
Editor: J. L. Vincent
1984. 21 figures, 18 tables. XIII, 190 pages
Soft cover DM 52,-. ISBN 3-540-13412-3

Springer-Verlag
Berlin
Heidelberg
New York
Tokyo